ENERGY SECTOR STANDARD OF THE PEOPLE'S REPUBLIC OF CHINA

中华人民共和国能源行业标准

Quality Degree Evaluate Standard of Unit Works in Hydropower Projects Part 2: Installation of Hydraulic Steel Structure and Hoists

水电工程单元工程质量等级评定标准
第 2 部分：金属结构及启闭机安装工程

NB/T 35097.2-2017

Replace SDJ 249.2-88

Chief Development Department: China Renewable Energy Engineering Institute

Approval Department: National Energy Administration of the People's Republic of China

Implementation Date: March 1, 2018

China Water & Power Press

中国水利水电出版社

Beijing 2024

All rights reserved. No part of this publication may be reproduced, stored in a retrieval system, or transmitted in any form or by any means—electronic, mechanical, photocopying, recording or otherwise, without prior written permission of the publisher.

图书在版编目（CIP）数据

水电工程单元工程质量等级评定标准 ：第2部分 ：金属结构及启闭机安装工程 ：NB/T 35097.2-2017 = Quality Degree Evaluate Standard of Unit Works in Hydropower Projects Part 2: Installation of Hydraulic Steel Structure and Hoists (NB/T 35097.2-2017) ：英文 / 国家能源局发布. -- 北京 ：中国水利水电出版社，2024. 7. -- ISBN 978-7-5226-2645-1

Ⅰ. TV512-65

中国国家版本馆CIP数据核字第2024AH5489号

ENERGY SECTOR STANDARD
OF THE PEOPLE'S REPUBLIC OF CHINA
中华人民共和国能源行业标准

Quality Degree Evaluate Standard of Unit
Works in Hydropower Projects
Part 2: Installation of Hydraulic Steel
Structure and Hoists

水电工程单元工程质量等级评定标准
第2部分：金属结构及启闭机安装工程

NB/T 35097.2-2017

Replace SDJ 249.2-88

（英文版）

Issued by National Energy Administration of the People's Republic of China
国家能源局　发布
Translation organized by China Renewable Energy Engineering Institute
水电水利规划设计总院　组织翻译
Published by China Water & Power Press
中国水利水电出版社　出版发行
　Tel: (+ 86 10) 68545888　68545874
　sales@mwr.gov.cn
　Account name: China Water & Power Press
　Address: No.1, Yuyuantan Nanlu, Haidian District, Beijing 100038, China
　http : //www.waterpub.com.cn
中国水利水电出版社微机排版中心　排版
北京中献拓方科技发展有限公司　印刷
184mm×260mm　16 开本　5.75 印张　182 千字
2024 年 7 月第 1 版　2024 年 7 月第 1 次印刷

Price（定价）：￥950.00

Introduction

This English version is one of China's energy sector standard series in English. Its translation was organized by China Renewable Energy Engineering Institute authorized by National Energy Administration of the People's Republic of China in compliance with relevant procedures and stipulations. This English version was issued by National Energy Administration of the People's Republic of China in Announcement [2022] No. 5 dated November 4, 2022.

This version was translated from the Chinese standard NB/T 35097.2-2017, *Quality Degree Evaluate Standard of Unit Works in Hydropower Projects—Part 2: Installation of Hydraulic Steel Structure and Hoists*, published by China Water & Power Press. The copy right is reserved by National Energy Administration of the People's Republic of China. In the event of any discrepancy in the implementation, the Chinese version shall prevail.

Many thanks go to the staff from the relevant standard development organizations and those who have provided generous assistance in the translation and review process.

For further improvement of the English version, any comments and suggestions are welcome and should be addressed to:

China Renewable Energy Engineering Institute
No. 2 Beixiaojie, Liupukang, Xicheng District, Beijing 100120, China
Website: www.creei.cn

Translating organization:

Sinohydro Engineering Bureau 8 Co., Ltd.

Translating staff:

| YAO Xianglian | FANG Zexia | WANG Huan | DENG Xiaoli |

Review panel members:

QIE Chunsheng	Senior English Translator
YAN Wenjun	Army Academy of Armored Forces, PLA
ZHANG Ming	Tsinghua University
GUO Jie	POWERCHINA Beijing Engineering Corporation Limited
JIA Haibo	POWERCHINA Kunming Engineering Corporation Limited

ZHANG Qingjun   China Gezhouba Group Machinery & Ship Corporation Limited

FAN Yilin   Sinohydro Engineering Bureau 8 Co., Ltd.

National Energy Administration of the People's Republic of China

翻译出版说明

本译本为国家能源局委托水电水利规划设计总院按照有关程序和规定，统一组织翻译的能源行业标准英文版系列译本之一。2022年11月4日，国家能源局以2022年第5号公告予以公布。

本译本是根据中国水利水电出版社出版的《水电工程单元工程质量等级评定标准 第2部分：金属结构及启闭机安装工程》NB/T 35097.2—2017翻译的，著作权归国家能源局所有。在使用过程中，如出现异议，以中文版为准。

本译本在翻译和审核过程中，本规范编制单位及编制组有关成员给予了积极协助。为不断提高本译本的质量，欢迎使用者提出意见和建议，并反馈给水电水利规划设计总院。

地址：北京市西城区六铺炕北小街2号
邮编：100120
网址：www.creei.cn

本译本翻译单位：中国水利水电第八工程局有限公司
本译本翻译人员：姚湘联　方泽霞　王　欢　邓小利
本译本审核人员：

  郄春生　英语高级翻译

  闫文军　中国人民解放军陆军装甲兵学院

  张　明　清华大学

  郭　洁　中国电建集团北京勘测设计研究院有限公司

  贾海波　中国电建集团昆明勘测设计研究院有限公司

  张庆军　中国葛洲坝集团机械船舶有限公司

  范一林　中国水利水电第八工程局有限公司

国家能源局

Announcement of National Energy Administration of the People's Republic of China [2017] No. 10

According to the requirements of Document GNJKJ [2009] No. 52, "Notice on Releasing the Energy Sector Standardization Administration Regulations (*tentative*) and detailed implementation rules issued by National Energy Administration of the People's Republic of China", 204 sector standards such as *Safety Management Specification of Coalbed Methane Production Station*, including 62 energy standards (NB), 86 electric power standards (DL), and 56 petroleum standards (SY), are issued by National Energy Administration of the People's Republic of China after due review and approval.

Attachment: Directory of Sector Standards

National Energy Administration of the People's Republic of China

November 15, 2017

Attachment:

Directory of Sector Standards

Serial number	Standard No.	Title	Replaced standard No.	Adopted international standard No.	Approval date	Implementation date
...						
31	NB/T 35097.2-2017	Quality Degree Evaluate Standard of Unit Works in Hydropower Projects Part 2: Installation of Hydraulic Steel Structure and Hoists	SDJ 249.2-88		2017-11-15	2018-03-01
...						

Foreword

According to the requirements of Document FGBGY [2008] No. 1242 "Notice on Printing and Publishing Sector Standard Project Plan in 2008" issued by National Development and Reform Commission Office, and after extensive investigation and study, summarization of practical experience and wide solicitation of opinions, the drafting group has prepared this standard.

The main technical contents of this standard are: project classification and quality evaluation, penstock manufacture, penstock installation, installation of plain gate, installation of radial gate, installation of miter gate, installation of trash rack, installation of fixed winch hoist, installation of mobile hoist, installation of hydraulic hoist and installation of screw rod hoist.

The main technical contents revised in this standard are as follows:

—revising the "excellent" evaluation criteria for unit works.

—adding the chapter "Project Classification and Quality Evaluation".

—revising the quality evaluation criteria for unit works.

—supplementing the quality evaluation criteria for the installation of caterpillar gate.

—supplementing the quality evaluation criteria for the installation of eccentric hinge compressed or pressure-actuated radial gate.

—deleting the chapter "Bridge Type Hoist Installation Work".

—deleting the section "Hoist Track Installation Work".

—deleting the "Appendix 1 Penstock Manufacture Works".

National Energy Administration of the People's Republic of China is in charge of the administration of this specification. China Renewable Energy Engineering Institute has proposed this specification and is responsible for its routine management. Energy Sector Standardization Technical Committee on Hydropower Steel Structures and Hoists is responsible for the explanation of specific technical contents. Comments and suggestions in the implementation of this specification should be addressed to:

China Renewable Energy Engineering Institute
No. 2, Beixiaojie, Liupukang, Xicheng District, Beijing, 100120, China

Chief drafting organization of this standard:

Sinohydro Engineering Bureau 8 Co., Ltd.

Participating drafting organizations of this standard:

Mechanical and Electrical Company of Sinohydro Engineering Bureau 8 Co., Ltd.

Tianjin Institute of Hydroelectric and Power Research

Chief drafting staff:

FAN Yilin	WANG Qimao	ZENG Hui	WANG Jianwen
HE Yinghong	DU Yi	LI Dongsheng	ZHANG Jiang
ZHANG Sheng			

Review panel members:

LIN Zhaohui	CHEN Hong	HU Baowen	FANG Hanmei
YAO Changjie	ZHANG Weiming	ZHAO Yongping	LI Lili
WAN Tianming	LUO Wenqiang	JIN Xiaohua	LIAO Yongping
YU Xiquan	LONG Zhaohui	YANG Qinghua	LI Liu
LI Shisheng			

Contents

1	**General Provisions**	1
2	**Terms and Symbols**	2
2.1	Terms	2
2.2	Symbols	2
3	**Project Classification and Quality Evaluation**	4
3.1	Project Classification	4
3.2	Quality Evaluation	6
4	**Penstock Manufacture**	8
4.1	Quality Evaluation Conditions	8
4.2	Inspection Items and Evaluation Criteria	8
5	**Penstock Installation**	20
5.1	Quality Evaluation Conditions	20
5.2	Evaluation of Unit Works for Installation of Embedded Penstock	20
5.3	Evaluation of Unit Works for Installation of Exposed Penstock	23
6	**Installation of Plain Gate**	25
6.1	Quality Evaluation Conditions	25
6.2	Evaluation of Unit Works for Installation of Embedded Parts	25
6.3	Evaluation of Unit Works for Installation of Gate Leaf	25
7	**Installation of Radial Gate**	39
7.1	Quality Evaluation Conditions	39
7.2	Evaluation of Unit Works for Installation of Embedded Parts	39
7.3	Evaluation of Unit Works for Installation of Gate Leaf	39
8	**Installation of Miter Gate**	47
8.1	Quality Evaluation Conditions	47
8.2	Evaluation of Unit Works for Installation of Embedded Parts	47
8.3	Evaluation of Unit Works for Installation of Gate Leaf	48
9	**Installation of Trash Rack**	51
9.1	Quality Evaluation Conditions	51
9.2	Evaluation of Unit Works for Installation of Embedded Parts and Trash Rack Bars	51
10	**Installation of Fixed Winch Hoist**	53
10.1	Quality Evaluation Conditions	53

10.2	Evaluation of Unit Works for Installation of Fixed Winch Hoist	53
11	**Installation of Mobile Hoist**	**56**
11.1	Quality Evaluation Conditions	56
11.2	Evaluation of Unit Works for Installation of Mobile Hoist	56
12	**Installation of Hydraulic Hoist**	**70**
12.1	Quality Evaluation Conditions	70
12.2	Evaluation of Unit Works for Installation of Hydraulic Hoist	70
13	**Installation of Screw Rod Hoist**	**73**
13.1	Quality Evaluation Conditions	73
13.2	Evaluation of Unit Works for Installation of Screw Rod Hoist	73
Appendix A	**Quality Evaluation Forms for Unit Works**	**75**
Explanation of Wording in This Standard		**77**
List of Quoted Standards		**78**

1 General Provisions

1.0.1 This standard is formulated with a view to standardizing the quality evaluation of unit works for the installation of hydraulic steel structure and hoists in hydropower projects to ensure quality, safety and reliability, technological advances and economy.

1.0.2 This standard is applicable to the quality evaluation of unit works for the installation of steel gate, trash rack and hoist, and the manufacture and installation of penstock in hydropower projects.

1.0.3 The quality evaluation of unit works shall be carried out after the inspection items have passed the acceptance in accordance with applicable acceptance specifications and have complete, full and accurate construction records.

1.0.4 In addition to this standard, the quality evaluation of the unit works for the installation of hydraulic steel structure and hoists in hydropower projects shall comply with other current standards of China.

2 Terms and Symbols

2.1 Terms

2.1.1 unit works

the smallest complex completed by one or several types of work for the installation of hydraulic steel structure and hoists in hydropower projects divided according to the design structure, construction process and quality evaluation requirements, which is the basic unit for construction quality evaluation

2.1.2 partitional project

installation works which perform a certain function in a structure or a building

2.1.3 unit project

works that consist of several partitional projects and have independent construction conditions or independent functions

2.1.4 main items

inspection items having a significant effect on the structural safety, service function and environment of unit works, which are marked with Δ in the evaluation forms of this standard

2.1.5 general items

inspection items other than main items

2.1.6 qualified items

measured values of 100 % of main items meeting the quatified criteria measured values of over 90 % of general items meeting the qualified criteria and that of the rest not greater than 1.2 times of the permissible deviation

2.1.7 excellent items

qualified items that more than 90 % of the measured points reach the excellent grade

2.2 Symbols

$P\,S_a$—te surface treatment grade of local blast cleaning for site construction;

R_z—the surface roughness, indicating the sum of the mean of five maximum contour peak heights and that of five maximum contour valley depths within the sampling length;

S_a—the grade of derusting through blasting or pigging for manufacture;

δ—the plate thickness or pipe wall thickness;

Δ—the main item.

3 Project Classification and Quality Evaluation

3.1 Project Classification

3.1.1 The hydropower project should be classified into unit works, partitional project and unit projects for quality evaluation.

3.1.2 The work breakdown for quality evaluation of the installation of hydraulic steel structure and hoists should be in accordance with Table 3.1.2.

Table 3.1.2 Work breakdown for quality evaluation of the installation of hydraulic steel structure and hoists

Unit project	partitional project	Unit works
Spillway works	Installation of gate and hoist	Installation of plain gate embedded parts
		Installation of plain gate leaf
		Installation of radial gate embedded parts
		Installation of radial gate leaf
		Installation of fixed winch hoist
		Installation of hydraulic hoist
		Installation of screw rod hoist
		Installation of mobile hoist
Flood discharge tunnel (outlet) works	Installation of gate and hoist	Installation of plain gate embedded parts
		Installation of plain gate leaf
		Installation of radial gate embedded parts
		Installation of radial gate leaf
		Installation of fixed winch hoist
		Installation of hydraulic hoist
		Installation of screw rod hoist
		Installation of mobile hoist
Overflow dam (outlet) works	Installation of gate and hoist	Installation of plain gate embedded parts
		Installation of plain gate leaf
		Installation of radial gate embedded parts
		Installation of radial gate leaf
		Installation of fixed winch hoist
		Installation of screw rod hoist
		Installation of hydraulic hoist
		Installation of mobile hoist

Table 3.1.2 (continued)

Unit project	partitional project	Unit works
Water conveyance works	Installation of gate and hoist, installation and manufacture of penstock, and installation of trash rack	Installation of plain gate embedded parts
		Installation of plain gate leaf
		Installation of hydraulic hoist
		Installation of mobile hoist
		Installation of trash rack embedded parts
		Installation of trash rack bar
		Manufacture of a single section or a pipe segment; installation of a pipe in a concrete-poured section or a part, such as lower elbow, inclined pipe or lower horizontal pipe
Headrace tunnel works	Installation of gate and hoist, installation and manufacture of penstock, and installation of trash rack	Installation of plain gate embedded parts
		Installation of plain gate leaf
		Installation of hydraulic hoist
		Installation of mobile hoist
		Installation of fixed winch hoist
		Installation of trash rack embedded parts
		Installation of trash rack bar
		Manufacture of a single section or a segment of steel pipe; installation of a pipe in a concrete-poured section or in a part, such as lower elbow, inclined pipe and lower horizontal pipe
Powerhouse works	Installation of gate and hoist	Installation of plain gate embedded parts
		Installation of plain gate leaf
		Installation of fixed winch hoist
		Installation of hydraulic hoist
		Installation of mobile hoist
Shiplock works	Installation of gate and hoist	Installation of plain gate embedded parts
		Installation of plain gate leaf
		Installation of radial gate embedded parts
		Installation of radial gate leaf
		Installation of miter gate embedded parts
		Installation of miter gate leaf
		Installation of fixed winch hoist

Table 3.1.2 *(continued)*

Unit project	partitional project	Unit works
Shiplock works	Installation of gate and hoist	Installation of hydraulic hoist
		Installation of mobile hoist
Diversion works	Installation of gate and hoist	Installation of plain gate embedded parts
		Installation of plain gate leaf
		Installation of hydraulic hoist
		Installation of fixed winch hoist
Water intake works	Installation of gate, hoist and trash rack	Installation of plain gate embedded parts
		Installation of plain gate leaf
		Installation of hydraulic hoist
		Installation of fixed winch hoist
		Installation of mobile hoist
		Installation of trash rack embedded parts
		Installation of trash rack bar

3.2 Quality Evaluation

3.2.1 The quality of the unit works is evaluated into two grades: qualified and excellent.

3.2.2 A unit works is evaluated as qualified if all the items are qualified.

3.2.3 A qualified unit works is evaluated as excellent if the excellent rates of the main and general items are above 70 %.

3.2.4 The following conditions shall be available before the quality evaluation of the unit works:

    1   All construction items of the unit works have been completed and accepted.

    2   All quality defects have been remedied or treatment opinions approved by the supervisor.

3.2.5 The quality evaluation of the unit works shall be carried out according to the following procedures:

    1   The contractor conducts self-inspection of the completed unit works, and applies to the supervisor for acceptance after the self-inspection passes.

    2   After receiving the application, the supervisor shall conduct spot

inspection and acceptance inspection. The construction quality of important concealed unit works and key part unit works shall be inspected and accepted by a team composed of the representatives from the owner, designer, supervisor, and contractor.

    3   The unit works shall be subjected to quality evaluation after passing the acceptance.

3.2.6 The quality evaluation of unit works shall include the following documents:

    1   Procedure documents such as construction drawings, construction schemes, construction records and acceptance documents.

    2   Quality inspection records and quality evaluation forms for unit works filled out by the contractor.

    3   Spot inspection documents with quality opinions signed by the supervising engineer.

3.2.7 The format of quality evaluation forms for unit works should be in accordance with Appendix A of this standard.

4 Penstock Manufacture

4.1 Quality Evaluation Conditions

4.1.1 Before delivery, penstocks shall be inspected and recorded according to the current national standard GB 50766, *Code for Manufacture Installation and Acceptance of Steel Penstocks in Hydroelectric and Hydraulic Engineering*, and be subjected to quality evaluation according to Section 3.2 of this standard.

4.1.2 The breakdown for quality evaluation of penstock manufacture shall meet the following requirements:

1. For straight, elbow and transition pipes, a section or a segment is regarded as a unit.
2. An expansion joint is regarded as a unit.
3. A bifurcation is regarded as a unit.

4.2 Inspection Items and Evaluation Criteria

4.2.1 The quality evaluation criteria for penstock circumference and section length shall be in accordance with Table 4.2.1.

Table 4.2.1 Quality evaluation criteria of penstock circumference and section length (mm)

No.	Item		Quality criteria		Remarks
			Qualified	Excellent	
$\Delta 1$	Difference between actual circumference and design circumference		$\pm 0.003D$, with absolute value ≤ 24	$\pm 0.0025D$, with absolute value ≤ 20	Installation circumferential seam without backing plate
			$\pm 0.003D$, whose absolute value ≤ 12	$\pm 0.0025D$, whose absolute value ≤ 10	Installation circumferential seam with V-shaped groove and backing plate, circumference at backing plate
$\Delta 2$	Circumference difference of adjacent pipe sections	$\delta < 10$	≤ 6	≤ 5	Installation circumferential seam without backing plate

Table 4.2.1 *(continued)*

No.	Item		Quality criteria		Remarks
			Qualified	Excellent	
Δ2	Circumference difference of adjacent pipe sections	$\delta < 10$	≤ 6	≤ 5	Installation circumferential seam with V-shaped groove and backing plate, circumference at backing plate
			≤ 10	≤ 9	Installation circumferential seam without backing plate
		$\delta \geq 10$	≤ 8	≤ 7	Installation circumferential seam with V-shaped groove and backing plate, circumference at backing plate
3	Difference between the length of a single-section penstock and its design length		≤ 5	≤ 4	–

NOTE: *D*—the inner diameter of the circular-section penstock (mm).

4.2.2 The quality evaluation criteria for planeness of penstock orifice shall comply with Table 4.2.2.

Table 4.2.2 Quality evaluation criteria for planeness of penstock orifice (mm)

No.	Item		Quality criteria	
			Qualified	Excellent
1	Planeness of penstock orifice	$D \leq 5000$	≤ 2	≤ 2
		$D > 5000$	≤ 3	≤ 2

4.2.3 The quality evaluation criteria for pipe cross-section geometry shall comply with Table 4.2.3.

Table 4.2.3 Quality evaluation criteria for pipe cross-section geometry (mm)

Cross-section geometry	Quality criteria		Remarks
	Qualified	Excellent	
Circular	Roundness ≤ 0.003D and ≤ 30	Roundness ≤ 0.003D and ≤ 25	Installation circumferential seam without backing plate
	When installing stiffener ring on penstock, the difference between the maximum and minimum diameters of the pipe mouth at the same end is not greater than 4	When installing stiffener ring on penstock, the difference between the maximum and minimum diameters of the pipe mouth at the same end is not greater than 3	Installation circumferential seam with V-shaped groove and backing plate, at least 4 pairs of diameters are measured at the pipe mouth of each end
Oval	The permissible deviation of major axis a and minor axis b of the oval is ±0.003a or ±0.003b, with absolute value is not greater than 6	The permissible deviation of major axis a and minor axis b of the oval is ±0.003a or ±0.003b, with absolute value not greater than 5	—
Rectangle	The permissible deviation of long side A and short side B is ±0.003A or ±0.003B, whose absolute value is not greater than 6 and the diagonal difference is not greater than 6	The permissible length deviation of long side A and short side B is ±0.003A or ±0.003B, whose absolute value is not greater than 5 and the diagonal difference is not greater than 5	—
Regular-polygon	The permissible deviation of circumference diameter is ±6; the difference between the maximum and minimum diameters is not greater than 0.003D, and not greater than 8	The permissible deviation of circumference diameter is ±6; the difference between the maximum and minimum diameters is not greater than 0.003D, and not greater than 7	—
Non-circular	The local planeness of penstock is not greater than 4 per meter	The local planeness of the penstock is not greater than 3 per meter	—

4.2.4 The quality evaluation criteria for the installation of support ring, stiffener ring, anti-thrust ring and dam ring shall comply with Table 4.2.4.

Table 4.2.4   Quality evaluation criteria for the installation of support ring, stiffener ring, anti-thrust ring and dam ring (mm)

No.	Item	Support ring		Stiffener ring, anti-thrust ring and dam ring		Sketch
		Qualified	Excellent	Qualified	Excellent	
1	Perpendicularity a of support ring or stiffener ring to pipe wall	$a \leq 0.01H$ and ≤ 3		$a \leq 0.02H$ and ≤ 5		
2	Perpendicularity b of the plane formed by support ring or stiffener ring to the pipe axis	$b \leq 0.002D$ and ≤ 6		$b \leq 0.004D$ and ≤ 12		
3	Permissible deviation of spacing C between two adjacent ring plates	±10		±30		eight points shall be measured for each circle

NOTE: For each item in the sketch, eight points shall be measured for each circle.

4.2.5 The quality evaluation criteria for the radial dislocation of longitudinal seam and circumferential seam shall comply with Table 4.2.5.

Table 4.2.5   Quality evaluation criteria for the radial dislocation of longitudinal seam and circumferential seam (mm)

No.	Item		Quality criteria	
			Qualified	Excellent
Δ1	Radial dislocation of longitudinal seam		0.1δ and ≤ 2.0	0.1δ and ≤ 1.5
2	Radial dislocation of circumferential seam	$\delta \leq 30$	0.15δ and ≤ 3.0	0.1δ and ≤ 2.0
		$30 < \delta \leq 60$	$\leq 0.1\delta$	$\leq 0.08\delta$
		$\delta > 60$	≤ 6.0	≤ 5.0

Table 4.2.5 *(continued)*

No.	Item	Quality criteria	
		Qualified	Excellent
Δ3	Dislocation of stainless steel clad plate weld	0.1δ and ≤ 1.5	0.1δ and ≤ 1.0

4.2.6 The quality evaluation criteria for the post-welding deformation of longitudinal seam shall comply with Table 4.2.6.

Table 4.2.6 Quality evaluation criteria for the post-welding deformation of longitudinal seam (mm)

No.	Item		Quality criteria		Chord length of template	Remarks
			Qualified	Excellent		
Δ1	Post-welding deformation of longitudinal seam	$D \leq 5000$	≤ 4	≤ 3	500	Installation circumferential seam without backing plate
			≤ 2	≤ 2		Installation circumferential seam with V-shaped groove and backing plate
		$5000 < D \leq 8000$	≤ 4	≤ 3	0.1D	Installation circumferential seam without backing plate
			≤ 2	≤ 2		Installation circumferential seam with V-shaped groove and backing plate
		$D > 8000$	≤ 6	≤ 4	1200	Installation circumferential seam without backing plate
			≤ 2	≤ 2		Installation circumferential seam with V-shaped groove and backing plate

4.2.7 The quality evaluation criteria for weld appearance shall comply with Table 4.2.7.

Table 4.2.7 Quality evaluation criteria for weld appearance (mm)

No.	Item	Quality criteria	
		Qualified	Excellent
Δ1	Crack	Not allowed for Classes Ⅰ, Ⅱ and Ⅲ welds	
2	Surface slag inclusion	Not allowed for Classes Ⅰ and Ⅱ welds; for Class Ⅲ weld, the slag inclusion depth is not greater than 0.1δ, and the length is not greater than 0.3δ and not greater than 10	
3	Undercut	For Classes Ⅰ and Ⅱ welds, the undercut depth shall not be greater than 0.5; for Class Ⅲ weld, the depth is not greater than 1	
4	Surface pore	Not allowed for Classes Ⅰ and Ⅱ welds; for Class Ⅲ weld, five pores, with a diameter less than 1.5 and spacing not less than 20, are allowed per meter	
Δ5	Incompletely filled weld	Not allowed for Classes Ⅰ and Ⅱ welds; for Class Ⅲ weld, it shall not be greater than $(0.2+0.02\delta)$ and not greater than 1; the total length of incompletely filled weld per 100 mm weld is not greater than 25 mm	
6	Weld reinforcement Δh	Manual welding	For Classes Ⅰ and Ⅱ welds: $\delta \leq 25, \Delta h = 0\text{-}2.5$ $25 < \delta \leq 50, \Delta h = 0\text{-}3$ $\delta > 50, \Delta h = 0\text{-}4$
		Automatic welding	For Classes Ⅰ and Ⅱ welds: 0-4 No requirement for Class Ⅲ weld
7	Butt weld width	Manual welding	Covering the groove on each side by 1 to 2.5, and transiting smoothly
		Automatic welding	Covering the groove on each side by 2 to 7, and transiting smoothly
Δ8	Fillet weld leg K	Where $K \leq 12$, K_{-1}^{+2}; where $K > 12$, K_{-1}^{+3}	
9	Spatter	Cleared for Classes Ⅰ and Ⅱ welds; no requirement for Class Ⅲ weld	
10	Weld beading	Not allowed for Classes Ⅰ and Ⅱ welds; no requirement for Class Ⅲ weld	

4.2.8 The evaluation criteria for the internal quality of Classes Ⅰ and Ⅱ welds shall comply with Table 4.2.8.

Table 4.2.8 Evaluation criteria for the internal quality of Classes I and II welds

No.	Item	Quality criteria	
		Qualified	Excellent
A1	Radiographic testing (RT)	The testing shall comply with the relevant provisions of the current national standard GB/T 3323, *Radiographic Examination of Fusion Welded Joints in Metallic Materials*, Unter the testing technology level of Level B, Class I welds are qualified if they are not lower than Level II, and Class II welds are qualified if they are not lower than Level III	One-time pass rate: 85 %
A2	Pulse reflection ultrasonic testing (UT) and phased array ultrasonic testing (PA-UT)	The testing shall comply with the current national standard GB/T 11345, *Non-destructive Testing of Welds—Ultrasonic Testing—Techniques, Testing Levels, and Assessment*, and the testing technology is of Level B. The quality level of welded joint shall be in accordance with the current national standard GB/T 29712, *Non-destructive Testing of Welds—Ultrasonic Testing—Acceptance Levels*. For Class I weld, it shall not lower than Level II; for Class II weld, it shall not be lower than Level III	One-time pass rate: 93 %
A3	Time of flight diffraction testing (TOFD)	The testing shall comply with the current sector standard DL/T 330, *Ultrasonic Time of Flight Diffraction Technique in Welded Joints of Hydroelectric or Hydraulic Engineering Steed Structure and Equipment*, or the sector standard NB/T 47013.10, *Nondestructive Testing of Pressure Equipments—Part 10: Ultrasonic Time of Flight Diffraction Technique*. For the quality grade of Classes I and II weld joint, it shall not be lower than Level II	One-time pass rate: 85 %
4	Magnetic particle testing (MT)	The testing shall comply with the current sector standard NB/T 47013.4, *Nondestructive Testing of Pressure Equipments—Part 4: Magnetic Particle Testing*. For Class I weld, it shall not be lower than Level II; for Class II weld, it shall not be lower than Level III	One-time pass rate: 95 %
5	Penetrant testing (PT)	The testing shall comply with the current sector standard NB/T 47013.5, *Nondestructive Testing of Pressure Equipments—Part 5: Penetrant Testing*; Class I weld is qualified when not lower than Level II, while Class II weld is qualified when not lower than Level III	One-time pass rate: 95 %

4.2.9 The quality evaluation criteria for the surface cleaning and local pit repair welding of the inner and outer walls of penstock shall comply with Table 4.2.9.

Table 4.2.9   Quality evaluation criteria for the surface cleaning and local pit repair welding of inner and outer walls of penstock

No.	Item	Quality criteria	
		Qualified	Excellent
1	Surface cleaning of the inner and outer walls of penstock	On the inner and outer walls, the temporary support, fixture and crater shall be cleared, and the wall surface shall be polished smoothly	
2	Local pit repair welding on the inner and outer walls of penstock	On the inner and outer walls, the pit with a depth greater than 0.1δ or greater than 2 mm shall be repaired by welding and polished to smooth	On the inner and outer walls, the pit with a depth greater than 0.1δ or greater than 1.5 mm shall be repaired by welding and polished to smooth

4.2.10 The quality evaluation criteria for the corrosion protection of penstock shall comply with Table 4.2.10.

Table 4.2.10   Quality evaluation criteria for corrosion protection of penstock

No.	Item	Quality criteria	
		Qualified	Excellent
1	Penstock inner wall surface treatment	The surface cleanliness of the inner wall of exposed or embedded penstock shall reach Grade Sa2.5 specified in the current national standard GB/T 8923.1, *Preparation of Steel Substrates Before Application of Paints and Related Products—Visual Assessment of Surface Cleanliness—Part 1: Rust Grades and Preparation Grades of Uncoated Steel Substrates and of Steel Substrates after Overall Removal of Previous Coatings*; the surface roughness of conventional coating is $R_z 40$ μm to $R_z 70$ μm, and that of the high-build heavy coating and metal hot spraying shall reach $R_z 60$ μm to $R_z 100$ μm	
2	Penstock outer wall surface treatment	For the spray coating on the outer wall of exposed penstock, the derusting grade is Sa2.5, and the surface roughness is $R_z 40$ μm to $R_z 70$ μm; for the modified cement paste and caustic soda cement paste spraying on the outer wall of embedded penstock, the derusting grade is Sa2	

Table 4.2.10 *(continued)*

No.	Item	Quality criteria	
		Qualified	Excellent
Δ3	Metal spraying visual inspection	The metal sprayed surface shall be uniform and free of sundry, peeling, bulging, holes, unevenness, loosely attached coarse metal particles, falling blocks, bare spots, and cracks	
Δ4	Metal spraying thickness and adhesion inspection	For all measuring points, the thickness is greater than 80 % of the design value, and the adhesion ensures that the coating is free of peeling from the base metal	
Δ5	Paint coating visual inspection	The coating surface shall be smooth, uniform in color, and free of wrinkle, blistering, sagging, pinholes, cracks, and holiday. The cement paste coating shall be of the same thickness basically, be adhered firmly and free of powdering	
Δ6	Internal quality inspection of coating	The coating thickness shall ensure that 85 % of the measuring points reach the design value. For the measuring point failing to reach the design thickness, the minimum thickness is not less than 85 % of the design thickness; When tested by a pinhole detector, the high-build coating shall be free of pinhole; The adhesion testing is carried out using the cross-cut method or pull method, and shall comply with the requirements of the design and the current national standard GB 50766, *Code for Manufacture Installation and Acceptance of Steel Penstocks in Hydroelectric and Hydraulic Engineering*	

4.2.11 The quality evaluation criteria for the manufacture of expansion joint shall comply with Table 4.2.11.

Table 4.2.11 Quality evaluation criteria for manufacture of expansion joint (mm)

No.	Item	Quality criteria	
		Qualified	Excellent
Δ1	Difference between actual circumference and design circumference of the water seal ring for inner and outer sleeves	±0.003D, with absolute value ≤ 8	±0.0025D, with absolute value ≤ 7

Table 4.2.11 *(continued)*

No.	Item	Quality criteria	
		Qualified	Excellent
Δ2	Clearance between post-welding radian of water seal ring and the template for inner and outer sleeves	< 2.0 at the longitudinal seam while < 1.0 at other positions	< 1.5 at the longitudinal seam while < 1.0 at other positions
Δ3	Difference between maximum/ minimum clearance and average clearance for inner and outer sleeves	The absolute value is not greater than 10 % of the average clearance	The absolute value is not greater than 8 % of the average clearance
4	Weld visual inspection	Comply with Table 4.2.7	
Δ5	Quality inspection for internal welding of Classes I and II welds	Comply with Table 4.2.8	
Δ6	Deviation between measured diameter and design diameter of water seal ring for inner and outer sleeves	±0.001D, whose absolute value ≤ 2.5	±0.001D, whose absolute value ≤ 2.0
Δ7	Air tightness test of bellow expansion joint	Comply with design requirements; when the water head is less than 25 m, only a kerosene penetrant test may be done	
8	Surface cleaning and local pit repair welding of pipe wall of water seal ring for inner and outer sleeves	Comply with Table 4.2.9	
9	Anti-corrosion surface treatment of inner and outer pipe walls of water seal ring for inner and outer sleeves	Comply with Table 4.2.10	
10	Coating of inner and outer pipe walls of water seal ring for inner and outer sleeves	Comply with Table 4.2.10	

4.2.12 The quality evaluation criteria for the manufacture of ribbed beam bifurcation shall comply with Table 4.2.12.

Table 4.2.12 Quality evaluation criteria for manufacture of ribbed beam bifurcation

No.	Item		Quality criteria	
			Qualified	Excellent
Sketch				
1	Pipe section lengths $L1$ and $L2$		±10	±8
2	Roundness of main and branch pipe mouths		$0.003D$ and ≤ 20	$0.003D$ and ≤ 18
Δ3	Difference between the measured circumference and design circumference of the main and branch pipe mouths		±0.003D, with a permissible deviation of ±20 and circumference difference of adjacent pipe joints not greater than 10	±0.003D, with a permissible deviation of ±18 and circumference difference of adjacent pipe joints not greater than 8
4	Center distance $S1$ between branch pipes		±10	±8
5	Dislocation of longitudinal seam		0.1δ and ≤ 2.0	0.1δ and ≤ 1.5
6	Dislocation of circumferential seam	$\delta \leq 30$	0.15δ and ≤ 3.0	0.1δ and ≤ 2.0
		$30 < \delta \leq 60$	0.1δ	0.08δ
		$\delta > 60$	≤ 6	≤ 6
7	Center height difference between main and branch pipes (D is the inner diameter of main pipe)	$D \leq 2000$	±4	±3
		$2000 < D \leq 5000$	±6	±5
		$D > 5000$	±8	±6

Table 4.2.12 *(continued)*

No.	Item		Quality criteria	
			Qualified	Excellent
8	Perpendicularity of main and branch pipe mouths	$D \leq 5000$	≤ 2.0	≤ 1.5
		$D > 5000$	≤ 3.0	≤ 2.5
9	Planeness of main and branch pipe mouths	$D \leq 5000$	≤ 2.0	≤ 1.5
		$D > 5000$	≤ 3.0	≤ 2.5
Δ10	Weld visual inspection		Comply with Table 4.2.7	
Δ11	Quality inspection for internal welding of Classes I and II welds		Comply with Table 4.2.8	
12	Surface cleaning and local pit repair welding of inner and outer pipe walls of bifurcation		Comply with Table 4.2.9	
13	Anti-corrosion surface treatment of bifurcation wall		Comply with Table 4.2.10	
14	Anti-corrosion metal spraying of bifurcation wall		Comply with Table 4.2.10	
15	Coating of bifurcation wall		Comply with Table 4.2.10	
Δ16	Water pressure test		Free of water seepage and other abnormal phenomena	

5 Penstock Installation

5.1 Quality Evaluation Conditions

5.1.1 The technical requirements for penstock installation shall comply with design drawings and the current national standard GB 50766, *Code for Manufacture Installation and Acceptance of Steel Penstocks in Hydroelectric and Hydraulic Engineering*.

5.1.2 For the quality degree evaluation of the unit works for penstock installation, the work breakdown shall be carried out in accordance with Article 4.1.2 of this standard.

5.1.3 The quality acceptance evaluation of the unit works for installation of penstock shall be in accordance with Section 3.2 of this standard. The evaluation shall include data such as the material certificate, re-measurement of main dimensions of pipe sections, testing record of installation quality inspection items, major defect remedy record, welding quality inspection record, surface corrosion protection record, water pressure test, and installation procedure documents.

5.2 Evaluation of Unit Works for Installation of Embedded Penstock

5.2.1 The quality evaluation criteria for embedded penstock mouth center and position shall be in accordance with Table 5.2.1.

5.2.2 The quality evaluation criteria for the penstock cross-section geometry shall be in accordance with Table 5.2.2.

5.2.3 The quality evaluation criteria for the dislocation of longitudinal seam and circumferential seam shall be in accordance with Table 4.2.5.

5.2.4 The quality evaluation criteria for weld appearance shall be in accordance with Table 4.2.7, and that of the internal quality of Classes I and II welds shall be in accordance with Table 4.2.8.

5.2.5 The quality evaluation criteria for the surface cleaning and local pit repair welding of the inner and outer walls of penstock shall be in accordance with Table 4.2.9.

5.2.6 For the anti-corrosion surface treatment for the installation weld of inner and outer walls of embedded penstock and its two sides, the quality evaluation criteria shall be in accordance with Table 5.2.6; for the metal spraying and paint coating, the quality evaluation shall be in accordance with Table 4.2.10.

Table 5.2.1　Quality evaluation criteria for embedded penstock mouth center and position (mm)

No.	Item	Quality criteria							
		Qualified				Excellent			
		Inner diameter D of penstock				Inner diameter D of penstock			
		$D \leq 2000$	$2000 < D \leq 5000$	$5000 < D \leq 8000$	$D > 8000$	$D \leq 2000$	$2000 < D \leq 5000$	$5000 < D \leq 8000$	$D > 8000$
Δ1	Orifice length deviation of initially installed section	±5	±5	±5	±5	±4	±4	±4	±4
Δ2	Orifice center deviation of initially installed section	±5	±5	±5	±5	±4	±4	±4	±4
Δ3	Perpendicularity deviation of penstock orifice on both ends of initially installed section	≤3	≤3	≤3	≤3	≤2	≤2	≤2	≤2
4	Orifice center deviation of penstock sections connected with spiral case, expansion joint, butterfly valve, ball valve and bifurcation as well as the start point of elbow	±6	±10	±12	±12	±5	±8	±10	±10
5	Center position deviation of other penstock sections in other positions	±15	±20	±25	±30	±10	±15	±20	±25
Δ6	Length deviation of start point of elbow	±10	±10	±10	±10	±8	±8	±8	±8

Table 5.2.2 Quality evaluation criteria for the penstock cross-section geometry (mm)

No.	Cross-section geometry	Quality criteria	
		Qualified	Excellent
1	Circular	The roundness is not greater than $0.005D$ and not greater than 40; at least two pairs of diameters are measured at each end of the penstock mouth	The roundness is not greater than $0.004D$ and not greater than 35; at least two pairs of diameters are measured at the penstock mouth at each end
2	Oval	The permissible length deviation of major axis a is $\pm 0.005a$, and that of the minor axis b is $\pm 0.005b$, whose absolute value is not greater than 8	The permissible length deviation of major axis a is $\pm 0.004a$, and that of the minor axis b is $\pm 0.004b$, whose absolute value is not greater than 6
3	Rectangle	The permissible length deviation of long side A is $\pm 0.005A$, while that of the short side B is $\pm 0.005B$, whose absolute value is not greater than 8 and the diagonal difference is not greater than 6	The permissible length deviation of long side A is $\pm 0.004A$, while that of the short side B is $\pm 0.004B$, whose absolute value is not greater than 6 and the diagonal difference is not greater than 4
4	Regular polygon	The permissible deviation of its circumference diameter is ± 8; the difference between the maximum diameter and the minimum diameter is not greater than $0.003D$ and not greater than 10	The permissible deviation of its circumference diameter is ± 6; the difference between the maximum diameter and the minimum diameter is not greater than $0.0025D$ and not greater than 8
5	Non-circular	The local planeness of penstock is not greater than 6 per meter	The local planeness of penstock is not greater than 5 per meter

Table 5.2.6 Quality evaluation criteria for anti-corrosion surface treatment for installation weld and its two sides

No.	Item	Quality criteria	
		Qualified	Excellent
Δ1	Anti-corrosion surface treatment of embedded penstock	The derusting grade of exposed face of embedded penstock reaches Grade P Sa2.5 specified in GB/T 8923.2, *Preparation of Steel Substrates Before Application of Paints and Related Products—Visual Assessment of Surface Cleanliness—Part 2: Preparation Grades of Previously Coated Steel Substrates after Localized Removal of Previous Coatings*. The surface roughness of conventional coatings reaches $R_z 40$ μm to $R_z 70$ μm, and that of high-build and metal hot spraying shall reach R_z 60 μm to $R_z 100$ μm. The surface cleanliness derusting grade of the part buried in concrete shall be Grade P Sa2	

5.2.7 The quality evaluation criteria for grout hole plug welding shall be in accordance with Table 5.2.7.

Table 5.2.7 Quality evaluation criteria for grout hole plug welding

No.	Item	Quality criteria		Remarks
		Qualified	Excellent	
1	Grout hole plug welding	The grout hole shall meet the design requirements after being welded, and the appearance of welded joint shall be subjected to MT or PT according to the current relevant sector standards NB/T 47013.4, *Nondestructive Testing of Pressure Equipments—Part 4: Magnetic Particle Testing* and NB/T 47013.5, *Nondestructive Testing of Pressure Equipment—Part 5: Penetrant Testing*, and the acceptance level shall be Level Ⅲ; the surface is flat and smooth without water seepage		For sampling ratio, PT or MT is adopted for grout hole: it is not less than 10 % for carbon steel and low-alloy steel, not less than 25 % for high-strength steel, and 100 % when cracks are found

5.3 Evaluation of Unit Works for Installation of Exposed Penstock

5.3.1 The mouth center and chainage quality of exposed penstock shall be evaluated as per Table 5.2.1.

5.3.2 The roundness quality evaluation of exposed penstock shall be in accordance with Table 5.2.2.

5.3.3 The quality evaluation of dislocation of longitudinal seam and circumferential seam shall be in accordance with Table 4.2.5.

5.3.4 The quality evaluation of weld appearance shall meet the requirements of Table 4.2.7.

5.3.5 The quality evaluation of internal quality of Classes I and II welds shall meet the requirements of Table 4.2.8.

5.3.6 The quality evaluation of surface cleaning and local pit repair welding shall meet the requirements of Table 4.2.9.

5.3.7 For the anti-corrosion surface treatment for the installation weld of inner and outer walls of exposed penstock and its two sides, the quality evaluation shall meet the requirements of Table 5.2.6; for the anti-corrosion metal spraying and paint coating, that shall meet the requirements of Table 4.2.10.

5.3.8 The quality evaluation criteria for center, elevation, radian and clearance of exposed penstock support shall meet the requirements of Table 5.3.8.

Table 5.3.8 Quality evaluation criteria for center, elevation, radian and clearance of exposed penstock support (mm)

No.	Item	Quality criteria	
		Qualified	Excellent
1	Permissible clearance between the top surface radian of saddle support and the template	≤ 2	≤ 2
Δ2	Elevation longitudinal center and transverse centers of abutment pad of rolling support or pendulum support	±5	±4
3	Parallelism with the design axis of penstock	≤ 2/1000	≤ 2/1000
4	Local clearance of contact surfaces of the rolling support or pendulum support	≤ 0.5	≤ 0.5
5	Longitudinal and horizontal inclinations of rolling, pendulum and sliding support abutment pads	≤ 2	≤ 2
6	Installation of rolling, pendulum and sliding supports	Operate agilely without jamming	

6 Installation of Plain Gate

6.1 Quality Evaluation Conditions

6.1.1 For embedded parts, the installation of a complete set of embedded parts or a section of them should be considered as one unit works; for gate leaf, the installation of each gate body should be considered as one unit works; the quality evaluation of unit works for installation of gate and gate slot shall be carried out according to the requirements of Section 3.2 of this standard.

6.1.2 The technical requirements for installation, welding, surface corrosion protection and inspection of embedded parts and gate leaf shall comply with the design drawings and the current sector standard NB/T 35045, *Code for Manufacture Installation and Acceptance of Steel Gates in Hydropower Engineering*.

6.1.3 For the quality acceptance evaluation of the unit works for installation of embedded parts and gate leaf, the installation procedures, installation records, welding and surface corrosion protection records, gate test and trial run, major defect remedy records, etc. shall be submitted.

6.2 Evaluation of Unit Works for Installation of Embedded Parts

6.2.1 The installation quality of the embedded parts of plain gate shall be evaluated as qualified when complying with Table 6.2.1-1, and be evaluated as excellent when complying with Table 6.2.1-2.

6.2.2 The installation quality of the embedded parts of caterpillar gate shall be in accordance with Table 6.2.1-1, Table 6.2.1-2 and Table 6.2.2.

6.2.3 The quality evaluation of the weld appearance of embedded parts shall be in accordance with Table 4.2.7.

6.2.4 For the anti-corrosion surface treatment for the installation weld of exposed penstock and its two sides, the quality evaluation shall be in accordance with Table 5.2.6; for the anti-corrosion metal spraying and paint coating, that shall be in accordance with Table 4.2.10.

6.3 Evaluation of Unit Works for Installation of Gate Leaf

6.3.1 The installation of plain gate includes that of plain sliding gate and fixed roller gate, and the quality evaluation shall be in accordance with Table 6.3.1.

Table 6.2.1-1 Quality evaluation criteria for installation of embedded parts of plain gate (mm)

No.	Name of embedded part		Bottom sill	Lintel	Main track		Side guide	Reverse guide	Seal plate	Corner guard serving as side guide	Breast wall			
					Machined	Unmachined					Serving as seal		Not serving as seal	
		Quality criteria									Upper	Lower	Upper	Lower
			Qualified	Qualified	Qualified	Qualified	Qualified	Qualified	Qualified	Qualified	Qualified	Qualified	Qualified	Qualified
1	For center line of gate slot, a	In working range	±5.0	+2.0 / −1.0	+2.0 / −1.0	+3.0 / −1.0	±5.0	+3.0 / −1.0	+2.0 / −1.0	±5.0	+5.0 / −0.0	+2.0 / −1.0	+8.0 / −0.0	+2.0 / −1.0
		Out of working range	—	—	+3.0 / −1.0	+5.0 / −2.0	±5.0	+5.0 / −2.0	—	±5.0	—	—	—	—
2	For center line of orifice, b	In working range	±5.0	—	±3.0	±3.0	±5.0	±3.0	±3.0	±5.0	—	—	—	—
		Out of working range	—	—	±4.0	±4.0	±5.0	±5.0	—	±5.0	—	—	—	—
3	Elevation		±5.0	—	—	—	—	—	—	—	—	—	—	—
Δ4	Distance between the lintel center and the bottom sill face, h		—	±3.0	—	—	—	—	—	—	—	—	—	—

Table 6.2.1-1 (*continued*)

No.	Quality criteria		Bottom sill	Lintel	Main track		Side guide	Reverse guide	Seal plate	Corner guard serving as side guide	Breast wall			
					Machined	Unmachined					Serving as seal		Not serving as seal	
											Upper	Lower	Upper	Lower
			Qualified	Qualified	Qualified	Qualified	Qualified	Qualified	Qualified	Qualified	Qualified	Qualified	Qualified	Qualified
Δ5	Height difference between two ends of working face	$L<10000$	≤2.0	–	–	–	–	–	–	–	–	–	–	–
		$L≥10000$	≤3.0	–	–	–	–	–	–	–	–	–	–	–
Δ6	Working face planeness	In working range	≤2.0	≤2.0	≤2.0	–	–	–	≤2.0	–	≤2.0	≤2.0	≤4.0	≤4.0
		Out of working range	–	–	–	–	–	–	–	–	–	–	–	–
Δ7	Dislocation at working face junction	In working range	≤1.0	≤0.5	≤0.5	≤1.0	≤1.0	≤1.0	≤0.5	≤1.0	≤1.0	≤1.0	≤1.0	≤1.0
		Out of working range	–	–	≤1.0	≤2.0	≤2.0	≤2.0	–	≤2.0	–	–	–	–

Table 6.2.1-1 *(continued)*

No.	Name of embedded part		Bottom sill	Lintel	Main track		Side guide	Reverse guide	Seal plate	Corner guard serving as side guide	Breast wall			
					Machined	Unmachined					Serving as seal		Not serving as seal	
											Upper	Lower	Upper	Lower
	Sketch										—			
8	Quality criteria		Qualified	Qualified	Qualified	Qualified	Qualified	Qualified	Qualified	Qualified	Qualified			
	Surface width in working range	B < 100	≤1.0	≤1.0	≤0.5	≤1.0	≤2.0	≤2.0	≤2.0	≤2.0	≤2.0			
		B = 100-200	≤1.5	≤1.5	≤1.0	≤2.0	≤2.5	≤2.5	≤2.5	≤2.5	≤2.5			
	Surface distortion, f	B > 200	≤2.0	—	≤1.0	≤2.0	≤3.0	≤3.0	≤3.0	—	≤3.0			
	Permissible increment if surface width is out of working range		—	—	≤2.0	≤2.0	≤2.0	≤2.0	≤2.0	—	≤2.0			

NOTES:
1 The lower part of the breast wall is the joint with the lintel.
2 The working height limit of gate slot: to the orifice height for gate opening and closing in static water; to the pressure-loaded main track height for gate opening and closing in dynamic water.
3 The dislocation of the joint shall be ground into a gentle slope.
4 At least one point shall be measured per meter.

Table 6.2.1-2 Excellent criteria for installation quality evaluation of embedded parts of plain gate (mm)

No.	Name of embedded part		Quality criteria	Bottom sill	Lintel	Main track Machined	Main track Unmachined	Side guide	Reverse guide	Seal plate	Corner guard serving as side guide	Breast wall Serving as seal Upper	Breast wall Serving as seal Lower	Breast wall Not serving as seal Upper	Breast wall Not serving as seal Lower
	Sketch														
				Excellent	Excellent	Excellent	Excellent	Excellent	Excellent	Excellent	Excellent	Excellent	Excellent	Excellent	Excellent
1	To center line of gate slot, a	In working range		±4.0	+1.0 / −0.5	+1.0 / −0.5	+2.0 / −1.0	±4.0	+2.0 / −1.0	+1.0 / −1.0	±4.0	+4.0 / −0.0	+1.0 / −1.0	+7.0 / −0.0	+1.0 / −1.0
		Out of working range		—	—	+2.0 / −1.0	+4.0 / −1.0	±4.0	+4.0 / −1.0	—	±4.0	—	—	—	—
2	To center line of orifice, b	In working range		±4.0	—	±2.0	±2.0	±4.0	±2.0	±2.0	±4.0	—	—	—	—
		Out of working range		—	—	±3.0	±3.0	±4.0	±4.0	—	—	—	—	—	—
3	Elevation			±4.0	—	—	—	—	—	—	—	—	—	—	—

Table 6.2.1-2 (continued)

No.	Name of embedded part	Bottom sill	Lintel	Main track		Side guide	Reverse guide	Seal plate	Corner guard serving as side guide	Breast wall			
				Machined	Unmachined					Serving as seal		Not serving as seal	
										Upper	Lower	Upper	Lower
	Sketch												
	Quality criteria	Excellent	Excellent	Excellent	Excellent	Excellent	Excellent	Excellent	Excellent	Excellent	Excellent	Excellent	Excellent
Δ4	Distance between the lintel center and the bottom sill face, h	—	±2.0	—	—	—	—	—	—	—	—	—	—
Δ5	Height difference between two ends of working face	$L<10000$	≤1.5	—	—	—	—	—	—	—	—	—	—	—
		$L≥10000$	≤2.0	—	—	—	—	—	—	—	—	—	—	—

Table 6.2.1-2 (continued)

No.	Name of embedded part	Bottom sill	Lintel	Main track					Corner guard serving as side guide	Breast wall				
				Machined	Unmachined	Side guide	Reverse guide	Seal plate		Serving as seal		Not serving as seal		
										Upper	Lower	Upper	Lower	
	Sketch													
	Quality criteria	Qualified	Qualified	Qualified	Qualified	Qualified	Qualified	Qualified	Qualified	Qualified	Qualified	Qualified	Qualified	
Δ6	Working face planeness — In working range	≤1.5	≤1.5	≤1.5	—	—	—	≤1.0	—	≤1.0	≤1.0	≤3.0	≤3.0	
	Out of working range	—	—	—	—	—	—	—	—	—	—	—	—	
Δ7	Dislocation at working face junction — In working range	≤0.5	≤0.5	≤0.5	≤0.5	≤0.5	≤0.5	≤0.5	≤0.5	≤0.5	≤0.5	≤0.5	≤0.5	
	Out of working range	—	—	≤0.5	≤1.5	≤1.0	≤1.0	—	≤1.0	—	—	—	—	

Table 6.2.1-2 (continued)

No.	Name of embedded part	Bottom sill	Lintel	Main track		Side guide	Reverse guide	Seal plate	Corner guard serving as side guide	Breast wall			
				Machined	Unmachined					Serving as seal		Not serving as seal	
										Upper	Lower	Upper	Lower
8	Sketch									—			
	Quality criteria	Excellent	Excellent	Excellent	Excellent	Excellent	Excellent	Excellent	Excellent	Excellent			
	Surface width in working range $B<100$	≤0.5	≤0.5	≤0.5	≤0.5	≤1.5	≤1.5	≤1.5	≤1.5	≤1.5			
	$B=100\text{-}200$	≤1.0	≤1.0	≤0.5	≤1.5	≤2.0	≤2.0	≤2.0	≤2.0	≤2.0			
	$B>200$	≤1.5	—	≤0.5	≤1.5	≤2.5	≤2.5	≤2.5	—	≤2.5			
	Permissible increment if surface width is out of working range	—	—	≤1.5	≤2.0	≤1.5	≤1.5	≤1.5	—	≤1.5			
	Surface distortion, f												

Table 6.2.2 Quality evaluation criteria for installation of embedded parts of caterpillar gate (mm)

No.	Testing item		Quality criteria	
			Qualified	Excellent
Δ1	Dislocation at the pressure-loaded surface joint of main track		≤ 0.2	≤ 0.1
Δ2	Planeness of the pressure-loaded surface of main track	$L ≤ 1000$	≤ 0.4	≤ 0.3
		$1000 < L ≤ 2500$	≤ 0.5	≤ 0.4
		$2500 < L ≤ 4000$	≤ 0.6	≤ 0.5
		$4000 < L ≤ 6300$	≤ 0.8	≤ 0.6
		$L > 6300$	≤ 1.0	≤ 0.8

NOTE: L—the main track length for embedded parts of caterpillar gate.

Table 6.3.1 Quality evaluation criteria for installation of plain gate (mm)

Sketch				
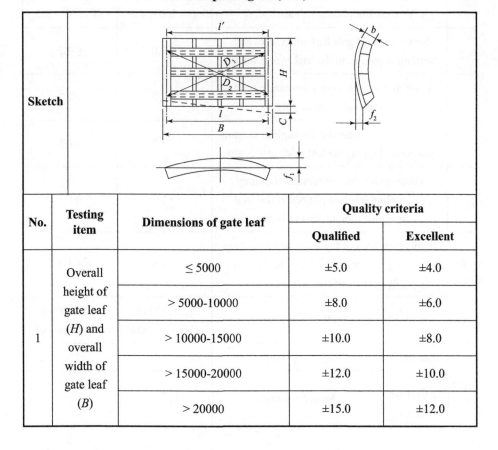				
No.	Testing item	Dimensions of gate leaf	Quality criteria	
			Qualified	Excellent
1	Overall height of gate leaf (H) and overall width of gate leaf (B)	≤ 5000	±5.0	±4.0
		> 5000–10000	±8.0	±6.0
		> 10000–15000	±10.0	±8.0
		> 15000–20000	±12.0	±10.0
		> 20000	±15.0	±12.0

Table 6.3.1 *(continued)*

No.	Testing item	Dimensions of gate leaf		Quality criteria	
				Qualified	Excellent
1	Difference between the corresponding sides of overall width of gate leaf (B) and those of overall height of gate leaf (H)			Not greater than half of the corresponding dimensional tolerance	
2	Diagonal relative Difference $|D_1 - D_2|$	The larger one of overall width of gate leaf (B) and overall height of gate leaf (H) shall prevail	≤ 5000	≤ 3.0	≤ 2.0
			> 5000-10000	≤ 4.0	≤ 3.0
			> 10000-15000	≤ 5.0	≤ 4.0
			> 15000-20000	≤ 6.0	≤ 5.0
			> 20000	≤ 7.0	≤ 6.0
3	Distortion		≤ 10000	≤ 3.0	≤ 2.0
			> 10000	≤ 4.0	≤ 3.0
4	Transverse straightness of gate leaf, f_1			$B/1500 ≤ 6.0$	$B/1500 ≤ 5.0$
5	Vertical straightness of gate leaf, f_2			$H/1500 ≤ 4.0$	$H/1500 ≤ 3.0$
6	Straightness of gate leaf when its overall bending is convex to the back water surface			≤ 3.0	≤ 3.0
7	Center distance between automatic coupling locating holes or pins			±2.0	±1.5
8	Distance from automatic decoupling locating pin center line to gate leaf thickness center			±2.0	±1.5
9	Distance between automatic coupling/ decoupling locating pin center line and specified longitudinal reference			±2.0	±1.5
10	Perpendicularity of the automatic coupling/ decoupling locating pin relative to the bottom edge planes of two side beams			≤ 1.5	≤ 1.5
11	Inclination of roller to any plane			≤ 2/1000 of wheel diameter	≤ 2/1000 of wheel diameter
Δ12	Planeness of roller or support slideway	Span $l ≤ 10000$		≤ 2.0	≤ 1.5
		Span $l > 10000$		≤ 3.0	≤ 2.0

Table 6.3.1 *(continued)*

No.	Testing item	Dimensions of gate leaf	Quality criteria			
			Qualified		Excellent	
13	Parallelism between support slideway and water seal seat	Slideway length: ≤ 500	≤ 0.5		≤ 0.5	
		Slideway length: > 500	≤ 1.0		≤ 1.0	
		Height difference of connection end between adjacent slideways	≤ 1.0		≤ 1.0	
14	Permissible span deviation of roller or support slideway	Span l	Roller	Slideway	Roller	Slideway
		$l ≤ 5000$	±2.0	±2.0	±2.0	±2.0
		$5000 < l ≤ 10000$	±3.0	±2.0	±3.0	±2.0
		$l > 10000$	±4.0	±2.0	±4.0	±2.0
15	Center line tolerance of roller or slideway on the same side		≤ 2.0		≤ 1.5	
Δ16	Permissible deviation of distance between the work face of the roller or support slideway and the water seal seat surface on the same cross section		±1.5		±1.0	
Δ17	Permissible deviation of distance between the work face of reverse support slideway or roller and the water seal seat surface		±2.0		±1.5	
Δ18	Permissible deviation of center distance of water seal on both sides		±3.0		±2.0	
19	Distance from the top water seal center to the bottom edge of bottom water seal		±3.0		±2.0	
Δ20	Planeness of rubber water seal		≤ 2.0		≤ 1.5	
Δ21	Difference between the actual compression amount and the design one of the rubber water seal		−1.0 to +2.0		−1.0 to +2.0	
22	Dislocation at gate assembly joint		≤ 2.0		≤ 1.5	

Table 6.3.1 (continued)

No.	Testing item Dimensions of gate leaf	Quality criteria	
		Qualified	Excellent
23	Clearance between installation positions in the case of gate assembly	≤ 4.0	≤ 3.0
Δ24	Static balance test of gate in the case of lifting	The inclination is neither greater than 1/1000 of the gate height nor greater than 8	The inclination is neither greater than 1/1000 of the gate height nor greater than 6
25	Stroke and tightness test of water filling valve	The stroke shall meet the design requirements, the guiding mechanism shall be flexible and reliable, the sealing element and seat valve shall contact uniformly and meet the water sealing requirements without light leakage and water leakage	
26	Permissible deviation of center of single lifting lug	±2.0	±1.5
	Permissible deviation of center distance between two lifting lugs of dual-lifting-point gate	±2.0	±1.5
	Permissible deviation of longitudinal and transverse center lines of lifting lug hole of gate	±2.0	±1.5
	Axle hole inclination of lifting lugs and hanger rods	≤ 1/1000	≤ 1/1000
27	Water-seal tightness inspection	In the case of working condition, the gate is inspected by light transmission or feeler gauge, and subjected to flushing test: water seal is qualified	

NOTE: B—overall width of gate leaf; H—overall height of gate leaf; l—span.

6.3.2 The installation quality of caterpillar gate shall be in accordance with Tables 6.3.1 and 6.3.2.

Table 6.3.2 Quality evaluation criteria for installation of caterpillar gate (mm)

No.	Testing item		Quality criteria	
			Qualified	Excellent
1	Overall height of gate leaf, H	$H \leq 10000$	±5.0	±4.0
		$10000 < H \leq 15000$	±7.0	±6.0
		$H > 15000$	±10	±8.0
2	Distortion of gate leaf	Overall width of gate leaf: $B \leq 10000$	≤3.0	≤2.0
		Overall width of gate leaf: $B > 10000$	≤4.0	≤3.0
Δ3	Water seal seat planeness		≤2.0	≤1.5
Δ4	Deviation of distance between water seal seat surface and downstream roller-chain surface		±0.8	±0.6
5	Planeness of opposed wheel		≤2.0	≤1.5
6	Deviation of distance between opposed wheel plane and roller-chain surface		±1.5	±1.0
7	Deviation of distance between opposed wheel center and non-bearing walkway center		±3.0	±2.5
8	Deviation of distance between side wheel rim surface and bearing walkway center		±3.0	±2.0
9	Clearance between side wheel and bearing plate surface of gate slot side wheel at any position		±5.0	±4.0
Δ10	Deviation of distance between the bottom surface of gate leaf and the water seal seat of gate slot		±1.0	±0.8
11	Bearing walkway span, L	$L \leq 5000$	±1.0	±0.8
		$5000 < L \leq 10000$	±2.0	±1.5
		$L > 10000$	±3.0	±2.5
12	Static balance test		≤ 1/1500 of gate height and ≤ 3.0	≤ 1/1500 of gate height and ≤ 2.5

Table 6.3.2 *(continued)*

No.	Testing item	Quality criteria	
		Qualified	Excellent
13	Sliding of the gate leaf in gate slot	Comply with the design requirements	
14	Assembly of gate leaf and gate slot	The chain and roller-chain shall operate freely without jamming and meet the design requirements	
Δ15	Contact between each sprocket and the bearing walkway surface when the gate leaf is placed horizontally	Contact well; the contact length is not less than 80 % of the sprocket length, and the local clearance is less than 0.10	
16	Lower sag between the roller-chain and the lower-end walkway when the gate leaf is in the working position	Comply with the design requirements	

6.3.3 The quality evaluation of weld appearance shall be in accordance with Table 4.2.7.

6.3.4 The quality evaluation criteria for internal quality of Classes I and II welds shall be in accordance with Table 4.2.8.

6.3.5 For the anti-corrosion surface treatment for the installation weld of gate leaf and its two sides, the quality evaluation shall be in accordance with Table 5.2.6; for the anti-corrosion metal spraying and paint coating, that shall be in accordance with Table 4.2.10.

6.3.6 The plain gate body shall be subjected to balance test, dry test, hydrostatic test and trial run according to the design requirements and relevant standards, and the records shall be made for future reference.

7 Installation of Radial Gate

7.1 Quality Evaluation Conditions

7.1.1 The work breakdown of the embedded parts and radial gate leaf shall be in accordance with Article 6.1.1 of this standard.

7.1.2 The technical requirements for the installation, welding, surface corrosion protection and inspection of the embedded parts and radial gate leaf shall be in accordance with Article 6.1.2 of this standard.

7.1.3 The quality acceptance evaluation of unit works for installation of embedded parts and radial gate leaf shall be in accordance with Article 6.1.3 of this standard.

7.2 Evaluation of Unit Works for Installation of Embedded Parts

7.2.1 The installation quality evaluation of the embedded parts of radial gate shall be in accordance with Tables 7.2.1-1 to 7.2.1-4.

7.2.2 For the pressure-actuated and eccentric hinge compressed radial gates, the quality evaluation criteria for installation shall be in accordance with Tables 7.2.1-1 to 7.2.1-4 and Table 7.2.2.

7.2.3 The quality evaluation criteria for weld appearance of embedded parts shall be in accordance with Table 4.2.7.

7.2.4 For installation weld of embedded parts and its two sides, the quality evaluation criteria for surface anti-corrosion treatment shall be in accordance with Table 5.2.6; and that for anti-corrosion metal spraying and paint coating shall be in accordance with Table 4.2.10.

7.3 Evaluation of Unit Works for Installation of Gate Leaf

7.3.1 The quality evaluation criteria for the installation of radial gate leaf shall be in accordance with Table 7.3.1.

7.3.2 The quality evaluation criteria for weld appearance shall be in accordance with Table 4.2.7.

7.3.3 The quality evaluation criteria for internal quality of Classes I and II welds shall meet the requirements of Table 4.2.8.

7.3.4 For the anti-corrosion surface treatment for the installation weld of gate leaf and its two sides, the quality evaluation criteria shall be in accordance with Table 5.2.6; for the anti-corrosion metal spraying and paint coating, that shall be in accordance with Table 4.2.10.

Table 7.2.1-1 Quality evaluation criteria I for installation of embedded parts of radial gate (mm)

No.	Name of embedded part		Bottom sill		Lintel		Side seal plate				Side wheel guide plate	
	Sketch						Submerged		Emersed			
	Quality criteria		Qualified	Excellent	Qualified	Excellent	Qualified	Excellent	Qualified	Excellent	Qualified	Excellent
1	Chainage		±5.0	±4.0	+2.0/-1.0	+1.0/-1.0	—	—	—	—	—	—
2	Elevation		±5.0	±4.0	—	—	—	—	—	—	—	—
Δ3	Distance between the threshold center and the bottom sill face, n		—	—	±3.0	±2.0	—	—	—	—	—	—
Δ4	To center line of orifice, b	In working range	±5.0	±4.0	—	—	±2.0	±1.5	+3.0/-2.0	+2.0/-1.0	+3.0/-2.0	+2.0/-1.0
		Out of working range	—	—	—	—	+4.0/-2.0	+3.0/-1.0	+6.0/-2.0	+5.0/-1.0	+6.0/-2.0	+5.0/-1.0
Δ5	Height difference between two ends of working face	$L < 10000$	≤2.0	≤1.0	—	—	—	—	—	—	—	—
		$L \geq 10000$	≤3.0	≤2.0	—	—	—	—	—	—	—	—
Δ6	Work face planeness		≤2.0	≤1.0	≤2.0	≤1.0	≤2.0	≤1.5	≤2.0	≤1.0	≤2.0	≤1.0
Δ7	Dislocation at working face junction		≤1.0	≤0.5	≤0.5	≤0.5	≤1.0	≤0.5	≤1.0	≤0.5	≤1.0	≤0.5
8	Curvature radii of center lines of side seal plate and side wheel guide plate		—	—	—	—	±5.0	±4.0	±5.0	±4.0	±5.0	±4.0

Table 7.2.1-1 *(continued)*

No.	Name of embedded part		Bottom sill		Lintel		Side seal plate				Side wheel guide	
	Sketch						Submerged		Emersed			
	Quality criteria		Qualified	Excellent	Qualified	Excellent	Qualified	Excellent	Qualified	Excellent	Qualified	Excellent
9	Surface distortion, f	Surface width in working range, B — $B<100$	≤1.0	≤0.5	≤1.0	≤0.5	≤1.0	≤0.5	≤1.0	≤0.5	≤2.0	≤1.5
		$100 \leq B \leq 200$	≤1.5	≤1.0	≤1.5	≤1.0	≤1.5	≤1.0	≤1.5	≤1.0	≤2.5	≤2.0
		$B>200$	≤2.0	≤1.5	—	—	≤2.0	≤1.5	≤2.0	≤1.5	≤3.0	≤2.5
	Permissible increment if surface width is out of working range		—	—	—	—	≤2.0	≤1.5	≤2.0	≤1.5	≤2.0	≤1.5

NOTES:
1 L—gate width.
2 Usually, the lintel is last fixed, and its position should be adjusted according to the actual position of gate leaf.
3 The working range refers to the water seal height.
4 At least one point on member is measured per meter.
5 For a submerged gate whose side seat plate is made of stainless steel, the deviation of dislocation at joint shall not be greater than 0.5 mm.
6 The dislocation at joint shall be ground into a gentle slope.

Table 7.2.1-2 Quality evaluation criteria II for installation of embedded parts of radial gate (mm)

No.	Item	Quality criteria			
		Submerged		Emersed	
		Qualified	Excellent	Qualified	Excellent
Δ1	Horizontal distance between the bottom sill center and the hinged support center	±4	±3	±5	±4
Δ2	Vertical distance between the bottom sill center and the hinged support center	±4	±3	±5	±4
Δ3	Distance between side seal seat plate center and hinged support center	±4	±3	±6	±5
4	Distance between water seal seat plates on both sides	+4 -3	+3 -2	+5 -3	+4 -2
5	Distance between wheel guides on both sides	+5 -3	+4 -2	+5 -3	+4 -2

Table 7.2.1-3 Quality evaluation criteria for installation of hinged support of steel radial gate (mm)

No.	Item	Quality criteria			
		Cylindrical hinge		Spherical hinge	
		Qualified	Excellent	Qualified	Excellent
1	Distance between hinged support center and orifice center	±1.5	±1.0	±1.5	±1.0
Δ2	Chainage of hinged support	±2.0	±1.5	±2.0	±1.5
Δ3	Elevation of hinged support	±2.0	±1.5	±2.0	±1.5
Δ4	Inclination of hinged support shaft hole	≤ $l/1000$	≤ $l/1000$	≤ $l/1000$	≤ $l/1000$
Δ5	Coaxiality of two hinged support axis	≤ 1.0	≤ 0.5	≤ 2.0	≤ 1.0
6	Foundation bolt center of hinged support	≤ 1.0	≤ 1.0	≤ 1.0	≤ 1.0

NOTE: The inclination of hinged support pin hole refers to the inclination in any direction, and l is the center distance between two ear plates.

Table 7.2.1-4 Quality evaluation criteria for installation of hinged support steel beam of radial steel gate (mm)

No.	Item	Quality criteria		Sketch
		Qualified	Excellent	
1	Distance between steel beam center and orifice center	±1.5	±1.0	
Δ2	Chainage of steel beam	±1.5	±1.5	
Δ3	Elevation of steel beam	±1.5	±1.5	
4	Inclination of hinged support shaft hole	≤ L/1000	≤ L/1000	

Table 7.2.2 Quality evaluation criteria for installation of embedded parts of pressure-actuated and eccentric hinge compressed radial gates (mm)

No.	Item	Quality criteria	
		Qualified	Excellent
Δ1	Distance between the main water seal seat base plane center line of embedded part and the orifice center line	±2.0	±1.5
Δ2	Curvature radius of the main water seal seat base plane of embedded part	±3.0, and the deviation direction shall be consistent with that of the curvature radius of the outer arc surface of gate leaf panel	±2.0, and the deviation direction shall be consistent with that of the curvature radius of the outer arc surface of gate leaf panel
3	Permissible dimension deviation of the clearance between the main water seal seat base plane of embedded part and the outer arc surface of radial gate	≤ 3.0, and the relative difference between the radius on both sides shall be not greater than 1.5	
4	Installation of seal clamp plate	The joint surface of the seal clamp plate and the water seal seat of gate frame shall be completely fitted, and the radius off-set joint of the seal clamp plate is less than 0.5 mm; the opening width of the outer portion of water seal head between the seal clamp plates is not less than the design value, and the distance between the top surface of seal clamp plate and the outer surface of gate slot or the top surface of the water seal head clamp plate shall comply with the design requirements	

Table 7.2.2 *(continued)*

No.	Item	Quality criteria	
		Qualified	Excellent
5	Installation of water seal	The bottom surface of seal clamp plate shall be in full contact with that of the water seal seat of gate slot and the water seal head shall not be higher than the seal clamp plate, which meets the design requirements	
6	Installation of main water seal for eccentric hinge compressed radial gate	Comply with the design requirements	

Table 7.3.1 Quality evaluation criteria for installation of radial gate leaf (mm)

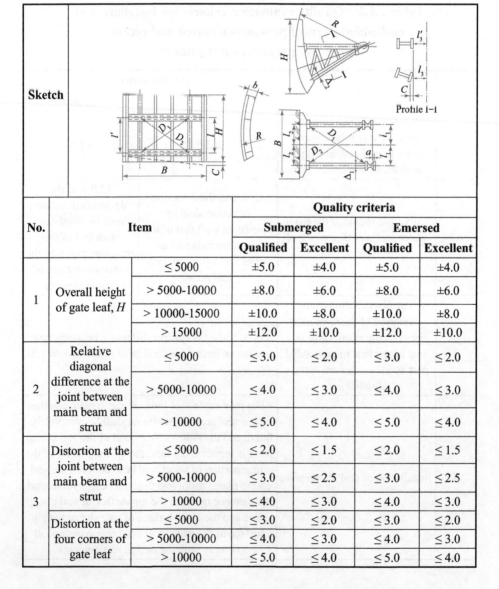

No.	Item		Quality criteria			
			Submerged		Emersed	
			Qualified	Excellent	Qualified	Excellent
1	Overall height of gate leaf, H	≤ 5000	±5.0	±4.0	±5.0	±4.0
		> 5000-10000	±8.0	±6.0	±8.0	±6.0
		> 10000-15000	±10.0	±8.0	±10.0	±8.0
		> 15000	±12.0	±10.0	±12.0	±10.0
2	Relative diagonal difference at the joint between main beam and strut	≤ 5000	≤ 3.0	≤ 2.0	≤ 3.0	≤ 2.0
		> 5000-10000	≤ 4.0	≤ 3.0	≤ 4.0	≤ 3.0
		> 10000	≤ 5.0	≤ 4.0	≤ 5.0	≤ 4.0
3	Distortion at the joint between main beam and strut	≤ 5000	≤ 2.0	≤ 1.5	≤ 2.0	≤ 1.5
		> 5000-10000	≤ 3.0	≤ 2.5	≤ 3.0	≤ 2.5
		> 10000	≤ 4.0	≤ 3.0	≤ 4.0	≤ 3.0
	Distortion at the four corners of gate leaf	≤ 5000	≤ 3.0	≤ 2.0	≤ 3.0	≤ 2.0
		> 5000-10000	≤ 4.0	≤ 3.0	≤ 4.0	≤ 3.0
		> 10000	≤ 5.0	≤ 4.0	≤ 5.0	≤ 4.0

Table 7.3.1 (continued)

No.	Item	Quality criteria			
		Submerged		Emersed	
		Qualified	Excellent	Qualified	Excellent
Δ4	Deviation of distance between hinge center and gate leaf center	±1.0	±0.8	±1.0	±0.8
5	Non-coincidence between arm column web center and main beam web center	≤ 4.0	≤ 3.0	≤ 4.0	≤ 3.0
Δ6	Coincidence between strut center and hinge center	≤ 2.0	≤ 1.5	≤ 2.0	≤ 1.5
Δ7	Deviation between strut center and gate leaf center	±1.5	±1.5	±1.5	±1.5
8	Curvature radius from hinge axis center to outer edge of panel, R	±4.0	±2.0	±8.0	±6.0
9	Relative difference between curvature radii of both sides	≤ 3.0	≤ 3.0	≤ 5.0	≤ 4.0
10	Eccentric hinge compressed or pressure-actuated radial gate	Curvature radius from hinge axis center to outer edge of panel, R	±3.0, and the deviation direction shall be consistent with that of the curvature radius of the water seal seat base plane of side guide	±2.0, and the deviation direction shall be consistent with that of the curvature radius of the water seal seat base plane of side guide	–	–
		Relative difference between curvature radii of both sides	≤ 1.5	≤ 1.0	–	–
		Deviation of the clearance between the main water seal seat base plane and the outer arc surface of radial gate	≤ 3.0	≤ 2.0	–	–
11	Difference between the actual compression amount and the design one of common radial gate rubber water seal	+2.0 −1.0	+1.0 −1.0	+2.0 −1.0	+1.0 −1.0

Table 7.3.1 *(continued)*

No.	Item	Quality criteria			
		Submerged		Emersed	
		Qualified	Excellent	Qualified	Excellent
12	Difference between the actual compression amount and the design one of the reverse radial gate top rubber water seal with the rigid water seal as bottom water seal	+2.0 -1.0	+2.0 -1.0	—	—
13	Difference between the actual compression amount and the design one of the reverse radial gate side rubber water seal with the rigid water seal as bottom water seal	0 -2.0	0 -2.0	—	—
14	Bottom water seal of reverse radial gate with the rigid water seal as bottom water seal contacts with the bottom sill	The local clearance, continuous length and cumulative length are not greater than 0.1 mm, 20 mm and 10 % of the full length respectively			
15	Joint surface between connecting plate at both ends of strut and hinge or main beam	Flat and close, with the contact surface not less than 75 %, after the connecting bolt is tightened, it is inspected with a 0.3 mm feeler gauge. The continuous insert part is not greater than 100 mm, and the cumulative length is not greater than 75 % of the circumference; the maximum clearance among a few points is not greater than 0.8 mm			
16	The shear plate and connecting plate at both ends of strut contact with each other	Contact tight			
17	Installation of water seal pressure charging and relief system equipment and pipeline for pressure-actuated radial gate	Comply with the design requirements			
18	Water pressure tightness test and water seal pressure charging test	Comply with the design requirements			
19	Water-seal tightness inspection	In the case of working condition, the gate is inspected by light transmission or feeler gauge, and subjected to flushing test: water seal is qualified			

7.3.5 The test and trial run of radial gate shall meet the requirements of design documents and relevant standards, and records shall be made for future reference.

8 Installation of Miter Gate

8.1 Quality Evaluation Conditions

8.1.1 The work breakdown of the embedded parts and miter gate leaf shall be in accordance with Article 6.1.1 of this standard.

8.1.2 The technical requirements for the installation, welding, surface corrosion protection and inspection of the embedded parts and miter gate leaf shall be in accordance with Article 6.1.2 of this standard.

8.1.3 The quality acceptance evaluation of the unit works for installation of embedded parts and miter gate leaf shall be in accordance with Article 6.1.3 of this standard.

8.2 Evaluation of Unit Works for Installation of Embedded Parts

8.2.1 The quality evaluation criteria for the installation of embedded parts of miter gates shall be in accordance with Section 8.1 of this standard.

8.2.2 The evaluation criteria for the unit works for installation of miter gate embedded parts shall also be in accordance with Table 8.2.2 of this standard.

Table 8.2.2 Quality evaluation criteria for installation of miter gate (mm)

No.		Testing item	Quality criteria	
			Qualified	Excellent
Δ1	Bottom pintle assembly	Semi-spherical pintle center position	±2.0	±1.5
2		Semi-spherical pintle elevation	±3.0	±2.0
Δ3		Relative height difference between two semi-spherical pintles	≤ 2.0	≤ 1.0
Δ4		Center distance between two semi-spherical pintles	±2.0	±1.5
Δ5		Horizontal inclination of bottom pintle assembly platform	≤ 1/1000	≤ 1/1250
Δ6	Top pintle assembly	Height difference between two ends of tie rod	≤ 1.0	≤ 0.8
Δ7		Coaxiality of the top pivot axis and the bottom one	≤ 2.0	≤ 1.5
Δ8		Deviation between the intersection point of the center lines of two tie rods and the top pivot center	≤ 2.0	≤ 1.5
Δ9		Perpendicularity of pin hole of tie rod seat	≤ 1/1000	≤ 1/1250

Table 8.2.2 *(continued)*

No.	Testing item		Quality criteria	
			Qualified	Excellent
10	Pillow seat	Deviation of pillow seat center line from support center	≤ 2.0	≤ 1.5
11	Embedded parts of bottom sill	Distance to the orifice center line	±2.0	±1.5
12		Distance to center line of gate axis	±2.0	±1.5
13		Planeness and straightness of work face	≤ 2.0	≤ 1.0
14		Dislocation at working face junction	≤ 0.5	≤ 0.5
15		Work face elevation of embedded parts	≤ 1.5	≤ 1.0

8.3 Evaluation of Unit Works for Installation of Gate Leaf

8.3.1 The quality evaluation criteria for the installation of miter gate leaf shall be in accordance with Table 8.3.1.

Table 8.3.1 Quality evaluation criteria for installation of miter gate leaf (mm)

Sketch	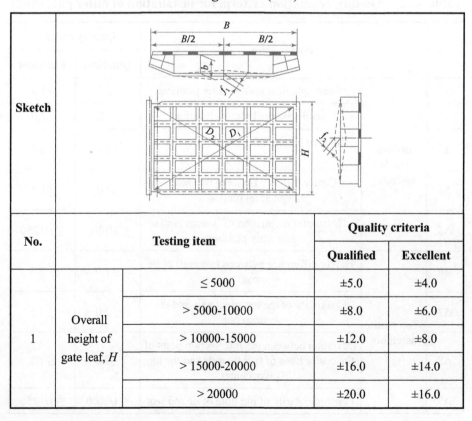			
No.	Testing item		Quality criteria	
			Qualified	Excellent
1	Overall height of gate leaf, H	≤ 5000	±5.0	±4.0
		> 5000-10000	±8.0	±6.0
		> 10000-15000	±12.0	±8.0
		> 15000-20000	±16.0	±14.0
		> 20000	±20.0	±16.0

Table 8.3.1 *(continued)*

No.	Testing item		Quality criteria	
			Qualified	Excellent
2	Relative diagonal difference, $|D_1 - D_2|$	≤ 5000	≤ 3.0	≤ 2.0
		> 5000-10000	≤ 4.0	≤ 3.0
		> 10000-15000	≤ 5.0	≤ 4.0
		> 15000-20000	≤ 6.0	≤ 5.0
		> 20000	≤ 7.0	≤ 6.0
Δ3	Concentricity between the intersection point of the center lines of two tie rods and the top pivot center		≤ 2.0	≤ 1.5
Δ4	Coaxiality between the top and bottom pintle assembly centers		≤ 2.0	≤ 1.5
5	Clearance between support and pillow seat spacers	Continuous clearance between support and pillow seat spacers for water seal	≤ 0.15	≤ 0.10
		Local clearance between support and pillow seat spacers for water seal	≤ 0.3	≤ 0.2
		Continuous clearance between support and pillow seat spacers not for water seal	≤ 0.2	≤ 0.1
		Local clearance between support and pillow seat spacers not for water seal	≤ 0.4	≤ 0.3
		Total clearance length	Not greater than 10 % of the total length of the support or pillow seat spacer	Not greater than 9 % of the total length of the support or pillow seat spacer
6	Center line offset of each pair of support and pillow seat spacers in contact with each other and not for water seal		≤ 5	≤ 4
7	Center line offset of each pair of support and pillow seat spacers in contact with each other and for water seal		≤ 3.0	≤ 2.5

Table 8.3.1 *(continued)*

No.	Testing item		Quality criteria	
			Qualified	Excellent
Δ8	Maximum runout at any point on miter post	Overall width B of gate leaf ≤ 12000	≤ 1.0	≤ 0.5
		Overall width B of gate leaf > 12000 but ≤ 24000	≤ 1.5	≤ 1.0
		Overall width B of gate leaf > 24000	≤ 2.0	≤ 1.5
Δ9	Sagging value of bottom beam of gate leaf at the miter post end		±2.0	±1.5
10	Relative design compression amount of side seal when closing the gate		+2.0 -1.0	+1.0 -1.0
11	Contact between the bottom water seal and the surface of embedded part when closing the gate		Uniform and free of clearance	
Δ12	Axis deviation values of top and bottom pintle assembly		≤ 2.0	≤ 1.5
13	Planeness of water seal surface		≤ 2.0	≤ 2.0

8.3.2 The quality evaluation criteria for weld appearance shall be in accordance with Table 4.2.7.

8.3.3 The quality evaluation criteria for internal welding of Classes Ⅰ and Ⅱ welds shall be in accordance with Table 4.2.8.

8.3.4 For gate leaf, the quality evaluation criteria for anti-corrosion surface treatment for installation weld and both sides of it shall be in accordance with Table 5.2.6, and that of anti-corrosion metal spraying and paint coating shall comply with those of Table 4.2.10.

8.3.5 The test and trial run of gate shall meet the design requirements and relevant standards, and records shall be made for future reference.

9 Installation of Trash Rack

9.1 Quality Evaluation Conditions

9.1.1 The work breakdown for embedded parts and trash rack bars shall be in accordance with Article 6.1.1 of this standard.

9.1.2 The technical requirements for installation, welding, surface corrosion protection and inspection of embedded parts and trash rack bars shall be in accordance with Article 6.1.2 of this standard.

9.1.3 The quality acceptance evaluation of unit works for installation of embedded parts and trash rack bars shall be in accordance with Article 6.1.3 of this standard.

9.2 Evaluation of Unit Works for Installation of Embedded Parts and Trash Rack Bars

9.2.1 The quality evaluation criteria for the installation of embedded parts of trash rack shall be in accordance with Table 9.2.1.

Table 9.2.1 Quality evaluation criteria for installation of embedded parts of trash rack

No.	Testing item	Quality criteria	
		Qualified	Excellent
1	Chainage of bottom sill, mm	±5.0	±4.0
2	Elevation of bottom sill, mm	±5.0	±4.0
3	Distance between center lines of bottom sill and orifice, mm	±5.0	±4.0
Δ4	Distance between center lines of main track and trash rack slot, mm	+3.0 / -2.0	+3.0 / -1.0
Δ5	Distance between center lines of reverse guide and trash rack slot, mm	+5.0 / -2.0	+4.0 / -1.0
6	Distance between center lines of main track/ reverse guide and orifice, mm	±5.0	±4.0
7	Inclination angle of inclined trash rack, '	≤ 10	≤ 10
8	Distance between work faces of main track and reverse guide, mm	+7.0 / -3.0	+5.0 / -2.0
9	Center distance of main track, mm	±8.0	±6.0
10	Center distance of reverse guide, mm	±8.0	±6.0

9.2.2 The quality evaluation criteria for the installation of trash rack bars shall be in accordance with Table 9.2.2.

Table 9.2.2 Quality evaluation criteria for installation of trash rack bars (mm)

No.	Item	Quality criteria	
		Qualified	Excellent
1	Height of trash rack	±8.0	±6.0
2	Relative diagonal difference of trash rack	≤ 6.0	≤ 5.0
3	Distortion	≤ 4.0	≤ 3.0
4	Span of slider or roller	±6.0	±5.0
5	Center line deviation of slider or roller support on the same side	±3.0	±2.0
6	Planeness of the slideway or roller work face of trash rack	≤ 4.0	≤ 3.0
7	Distance deviation of center line of lug hole of common trash rack	±4.0	±3.0
8	Deviation of lug center of trash rack when the hoist equipment is shared by trash rack and access door	±2.0	±1.5
Δ9	Connection between trash racks	Firm and reliable	
Δ10	Vertical movement of trash rack in trash rack slot	Agile, stable and free of jamming	
11	Trash rack with screen cleaner	The trash rack structure and the embedded parts of trash rack slot shall meet the operation requirements of the screen cleaner	

10 Installation of Fixed Winch Hoist

10.1 Quality Evaluation Conditions

10.1.1 The installation of each hoist should be regarded as a unit works.

10.1.2 The hoist shall be subjected to factory integral assembly and trial run before delivery, and shall pass the inspection. After arrival, the hoists shall be subjected to acceptance test according to the contract, and the main parts and components shall be retested, inspected and recorded.

10.1.3 The technical requirements for installation, welding, surface corrosion protection and inspection of hoist shall meet the requirements of design drawings and the current sector standard NB/T 35051, *Code for Manufacture Erection and Acceptance of Gate Hoists in Hydropower Projects*.

10.1.4 The assembly and testing in factory and the installation and trial run on site of hoist shall comply with the current sector standard NB/T 35051, *Code for Manufacture Erection and Acceptance of Gate Hoists in Hydropower Projects*.

10.2 Evaluation of Unit Works for Installation of Fixed Winch Hoist

10.2.1 The hoist shall move up and down in full stroke for 3 times at no load, and the inspection and adjustment of electrical and mechanical systems shall meet the requirements of the current sector standard NB/T 35051, *Code for Manufacture Erection and Acceptance of Gate Hoists in Hydropower Projects*.

10.2.2 For load test, the hoist shall first lift the gate in full stroke twice under conditions of no water in groove and static water; after the test passes, the open-close test under dynamic water condition of service gate and test of closing under dynamic water and opening under static water of emergency gate shall be carried out based on the design requirements according to the gate operating conditions, and the gates shall move vertically in full stroke twice; the quick-acting emergency gate shall be subjected to the closing test under dynamic water condition.

10.2.3 For load test, the inspection of electrical and mechanical systems of the hoist shall be in accordance with the current sector standard NB/T 35051, *Code for Manufacture Erection and Acceptance of Gate Hoists in Hydropower Projects*.

10.2.4 The quality evaluation criteria for the installation of fixed winch hoist shall be in accordance with Table 10.2.4.

Table 10.2.4 Quality evaluation criteria for installation of fixed winch hoist

No.	Testing item		Quality criteria	
			Qualified	Excellent
Δ1	Horizontal and longitudinal center lines of hoist body, mm		±3	±2.5
Δ2	Elevation of hoist body, mm		±5	±4
Δ3	Levelness of hoist body, mm		≤ 0.5/1000	≤ 0.4/1000
4	Distance between lifting points of dual-lifting-point hoist, mm		≤ 3.0	≤ 2.5
5	Height difference between center lines of two lifting shafts within the orifice part, mm		≤ 5	≤ 4
6	Height difference between center lines of two lifting shafts within full stroke of medium- and high-lift hoist, mm		≤ 30	≤ 24
7	Number of turns left on the drum when the lifting point is at the lower limit		Comply with the design requirements	
Δ8	Wire rope installation		Eliminate the winding stress before installation, and then assemble, wind the wire rope on the drum orderly and layer by layer without overlapping or crossing grooves, the rest parts shall not be subjected to flame cutting and lengthening, and shall comply with the design requirements	
Δ9	No-load test		Comply with Article 10.2.1 of this standard	
Δ10	Load test		Comply with Articles 10.2.2 and 10.2.3 of this standard	
Δ11	Temperature of rolling bearing	Temperature, °C	≤ 85	
		Temperature rise, K	≤ 35	
Δ12	Temperature of sliding bearing	Temperature, °C	≤ 70	
		Temperature rise, K	≤ 20	
13	Open gear pair contact spot		It shall not be less than 30 % in the tooth height direction, and shall not be less than 40 % accumulatively in the tooth length direction	
Δ14	Lubricating oil for rotating parts and components such as reducers, open gears, bearings, hydraulic brakes, etc.		Comply with the design requirements	

10.2.5 The quality evaluation criteria for unit works for installation of electrical equipment shall be in accordance with Table 10.2.5.

Table 10.2.5   Quality evaluation criteria for unit works for installation of electrical equipment

No.	Testing item		Quality criteria	
			Qualified	Excellent
Δ1	Installation perpendicularity of electrical panel and cabinet (per 1 m), mm		≤ 1.5	≤ 1.0
2	Horizontal installation deviation of electrical panel and cabinet, mm	Top of two adjacent panels	≤ 2.0	≤ 1.5
		Top of panels in a row	≤ 5.0	≤ 4.0
3	Installation deviation of panel surface of electrical panel and cabinet, mm	Top edges of two adjacent panels	≤ 1.0	≤ 1.0
		Panel dial in a row	≤ 5.0	≤ 4.0
4	Seam between installation panels of electrical panel and cabinet, mm		≤ 2.0	≤ 1.5
5	Deviation of bending degree of the cable sleeve after bending, which is not greater than the outer diameter D of the sleeve		10 %D	8 %D
Δ6	Insulation resistance of resistor at normal temperature, MΩ		≥ 1	
Δ7	Insulation resistance of a complete set of electrical equipment such as electrical panel, cabinet and connecting cable, MS	Primary and secondary circuits in non-humid area	≥ 1	
		Primary and secondary circuits in humid area	≥ 0.5	
Δ8	Insulation resistance at normal temperature of motors with a rated voltage below 1,000V, MΩ		≥ 0.5	
Δ9	Line insulation resistance, MΩ		≥ 1.0	
10	Current unbalance degree of three-phase stator of motor		≤ 10 %	
11	Temperature of electrical element		No abnormal heating	
12	Controller contact		No burning	

11 Installation of Mobile Hoist

11.1 Quality Evaluation Conditions

11.1.1 The work breakdown for installation of mobile hoist shall be in accordance with Article 10.1.1 of this standard.

11.1.2 Before delivery, the work on mobile hoist shall be in accordance with Article 10.1.2 of this standard.

11.1.3 The technical requirements for installation, welding, surface corrosion protection and inspection of mobile hoist shall be in accordance with Article 10.1.3 of this standard.

11.1.4 The assembly and testing in factory and the installation and trial run on site of mobile hoist equipment shall be in accordance with Article 10.1.4 of this standard.

11.2 Evaluation of Unit Works for Installation of Mobile Hoist

11.2.1 The no-load test, static load test and dynamic load test of mobile hoist shall comply with the current sector standard NB/T 35051, *Code for Manufacture Erection and Acceptance of Gate Hoists in Hydropower Projects*.

11.2.2 The quality evaluation criteria for the installation of crane track shall be in accordance with Table 11.2.2.

Table 11.2.2 Quality evaluation criteria for installation of crane track (mm)

No.	Item		Quality criteria	
			Qualified	Excellent
Δ1	Track center line deviation	$L \leq 10000$	≤ 2	≤ 1
		$L > 10000$	≤ 3	≤ 2
Δ2	Track gauge L	$L \leq 10000$	±3.0	±2.5
		$L > 10000$	±5	±4
Δ3	Longitudinal straightness of track		≤ 1.0/1500	≤ 0.8/1500
Δ4	Relative difference of elevations of two tracks on the same section	$L \leq 10000$	≤ 5	≤ 4
		$L > 10000$	≤ 8	≤ 6
5	Dislocation of the left, right and upper sides of track joint		≤ 1	≤ 0.5

Table 11.2.2 *(continued)*

No.	Item		Quality criteria	
			Qualified	Excellent
6	Track joint clearance	At 20 °C	≤ 2.0	≤ 1.5
		At 10 °C	≤ 1.5	≤ 1.0
7	Horizontal inclination value of track top		Not greater than 1/200 of the top width	

11.2.3 The quality evaluation criteria for installation of trolley track shall be in accordance with Table 11.2.3.

Table 11.2.3 Quality evaluation criteria for installation of trolley track (mm)

No.	Item			Quality criteria	
				Qualified	Excellent
Δ1	Distance between track center line deviation and track web center line, d (Figure 11.2.4-2)	Box beam with rails on one side	Where $\delta < 12$	$d \leq 6$	$d \leq 5$
			Where $\delta \geq 12$	$d \leq 0.5\delta$	$d \leq 0.5\delta$
		Single web or trussed beam		$d \leq 0.5\delta$	$d \leq 0.5\delta$
Δ2	Track gauge T (Figure 11.2.4-1)	$T \leq 2500$		±2.0	±1.5
		$T > 2500$		±3	±2
Δ3	Straightness of track center line	Box beam centered on track		≤ 3	≤ 2
		Box beam centered on track, with walking board, and bent to the platform side		≤ 3	≤ 2
Δ4	Relative difference of elevations of two tracks on the same section (Figure 11.2.4-1)	Track gauge $T \leq 2500$		≤ 3	≤ 2
		Track gauge $T > 2500$		≤ 5	≤ 4
5	Vertical and lateral dislocation at the top of the track joint			≤ 1.0	≤ 0.5

Table 11.2.3 *(continued)*

No.	Item			Quality criteria	
				Qualified	Excellent
6	Track joint clearance			≤ 2.0	≤ 1.5
7	Horizontal inclination value of track top surface			Not greater than 1/200 of the top surface width	
8	Local curvature of the track on the laying plane, f_0 (Figure 11.2.4-3)			Not greater than 1 in any 2 m range	
9	Local curvature f_0 of the track on the laying plane, in full length (Figure 11.2.4-4)	Where the track is laid in the middle of the box beam	Lateral deviation between the center of the trolley track and the theoretical center of the track	≤ 2.5	
		Where the track is laid on the inner side the box beam	Lateral deviation from the theoretical track center line	$f_1 \leq 4$ if outward, $f_2 \leq 1$ if inward	

11.2.4 The quality evaluation criteria for installation of mobile hoist shall be in accordance with Table 11.2.4.

Table 11.2.4 Quality evaluation criteria for installation of mobile hoist

No.	Testing item				Quality criteria	
					Qualified	Excellent
Δ1	Travel mechanism	Trolley span			±3.0 mm	±2.0 mm
		Relative difference of trolley spans	Trolley gauge $T \leq$ 2.5 m		≤ 2.0 mm	≤ 1.5 mm
			Trolley gauge $T >$ 2.5 m		≤ 3.0 mm	≤ 2.0 mm
		Crane span of bridge crane, S (Figure 11.2.4-6)	$L \leq 10$ m		±3.0 mm	±2.0 mm
			10 m $< L \leq 26$ m		±5.0 mm	±4.0 mm
			$L > 26$ m		±8.0 mm	±7.0 mm
		Relative difference of crane spans of bridge crane	$L \leq 10$ m		≤ 3.0 mm	≤ 2.0 mm
			10 m $< L \leq 26$ m		≤ 5.0 mm	≤ 4.0 mm
			$L > 26$ m		≤ 8.0 mm	≤ 7.0 mm

Table 11.2.4 *(continued)*

No.	Testing item			Quality criteria	
				Qualified	Excellent
Δ1		Crane span of gantry crane, S (Figure 11.2.4-7)	L ≤ 10 m	±5.0 mm	±4.0 mm
			10 m < L ≤ 26 m	±8.0 mm	±7.0 mm
			L > 26 m	±10.0 mm	±8.0 mm
		Relative difference of crane spans of gantry crane	L ≤ 10 m	≤ 5.0 mm	≤ 4.0 mm
			10 m < L ≤ 26 m	≤ 8.0 mm	≤ 7.0 mm
			L > 26 m	≤ 10.0 mm	≤ 8.0 mm
Δ2	Travel mechanism	Vertical inclination of wheel (Figure 11.2.4-8)		Not greater than $L/400$ of the measured length, and the lower wheel rim is tilted inward	
Δ3		Horizontal inclination of wheel (Figure 11.2.4-9)		Not greater than $L/1000$ of the measured length, and the inclination direction of the wheel on the same axis shall be reversed	
Δ4		Misalignment of wheels under the same end beam (Figure 11.2.4-10)	Two wheels under the same equalizing beam	≤ 1.0 mm	≤ 1.0 mm
			Adjacent wheels of adjacent equalizing beams	≤ 2.0 mm	≤ 1.5 mm
			Between other wheels	≤ 3.0 mm	≤ 2.0 mm
5		Maximum clearance between the driven wheel and the track top surface where the trolley driving wheel is in contact with the pre-assembled track (Figure 11.2.4-11)		Not greater than 1/600 of the gauge and not greater than 4 mm	Not greater than 1/600 of the gauge and not greater than 3 mm

Table 11.2.4 *(continued)*

No.	Testing item		Quality criteria	
			Qualified	Excellent
Δ6	Bridge and gantry assembly	Camber of the main beam (Figure 11.2.4-6 and Figure 11.2.4-7)	Comply with the design requirements, and the maximum camber is controlled within 1/10 of the bridge span in the middle span	
7		Warping of cantilever end of the main beam (Figure 11.2.4-7)	Comply with the design requirements	
8		Horizontal curvature of the main beam (Figure 11.2.4-6 and Figure 11.2.4-7)	Not greater than 1/2000 of the bridge or gantry span, and not greater than 20 mm	
Δ9		Relative diagonal difference of the upper structure of the bridge and gantry	≤ 5.0 mm	
10		Perpendicularity of the gantry leg in the span direction (Figure 11.2.4-3)	Not greater than 1/1000 the height of gantry leg, and the inclination directions shall be symmetrical to each other	
11		Relative height difference from the wheel work face to the flange surface on the leg	≤ 8.0 mm	≤ 7.0 mm
12		Relative diagonal difference of lower end plane and side vertical surface of the gantry leg (Figure 11.2.4-5) — Leg height $H_1 \leq 10$ m	≤ 10 mm	
		Relative diagonal difference of lower end plane and side vertical surface of the gantry leg (Figure 11.2.4-5) — Leg height $H_1 > 10$ m	≤ 15 mm	
13		Engaging dislocation of web and flange at the joint where a single piece flange is adopted for the connection between the gantry superstructure and the gantry leg	Not greater than 1/2 of the plate thickness	
14		Bridge and gantry assembly flange connection surface clearance	Local clearance is not greater than 0.3 mm, clearance area is not greater than 30 %, bolt connection is provided without clearance, and flange edge clearance is not greater than 0.8 mm	

Table 11.2.4 *(continued)*

No.	Testing item		Quality criteria	
			Qualified	Excellent
Δ15	Trial operation	No-load trial run	Comply with Article 11.2.1	
Δ16		Static load test	Comply with Article 11.2.1	
Δ17		Dynamic load test	Comply with Article 11.2.1	
18	Hoisting mechanism	Distance error of double lifting points	≤ 3.0 mm	≤ 2.5 mm
19		Height difference between center lines of the two lifting shafts in the orifice part	≤ 5.0 mm	≤ 4.0 mm
20		Height difference between center lines of the two lifting shafts within the full range	≤ 30 mm	≤ 25 mm
Δ21		Static balancing test	The longitudinal and horizontal inclination of the single-lifting-point hooking beam do not exceed 1/1000 of the beam height and not greater than 8 mm	
22	Installation of lifting beam	Assembly quality and inspection	Comply with the current sector standard NB/T 35051 and the design requirements	
23		Among hydraulic device, all sensors and wiring device	Sealed and waterproof; the cable socket shall not leak, and shall meet the design requirements	
24		Grade, viscosity and cleanliness of hydraulic oil in pump station	Comply with service environment and selected hydraulic elements	
25	Installation of cleaning grab bucket	Parts such as hinge point and supporting wheel after the final assembly	Rotate agilely, move accurately and reliably, without jamming and rubbing	

Table 11.2.4 (continued)

No.	Testing item		Quality criteria	
			Qualified	Excellent
26	Installation of cleaning grab bucket	Spacing deviation of harrow teeth	≤ 2.0 mm	≤ 2.0 mm
27		Straightness deviation of tooth tip	≤ 3.0 mm	≤ 3.0 mm
28		Relative diagonal difference of frame	≤ 4.0 mm	≤ 4.0 mm
Δ29		Distortion	≤ 2.0 mm	≤ 2.0 mm
30		Maximum midspan vertical deflection of the main beam at full load	Not greater than 1/2000 of the upper main beam span	
31		Misalignment of guide wheels at the same side	≤ 2.0 mm	≤ 1.5 mm
32		Span deviation of guide wheel	≤ 2.0 mm	≤ 1.5 mm
Δ33		Deviation of center distance between lifting points	≤ 2.0 mm	≤ 1.5 mm
34	Installation of lubricating device	Each single-point lubricated bearing during assembling	Inject a proper amount of clean lubricating grease	
		Oil transfer circuit	Smooth and free of swarf and dirt	
		All parts of the lubricating oil circuit before the test	Check one by one to ensure smooth flow and compliance with design requirements	
Δ35	Wire rope	Number of turns of the wire rope on the drum where the lifting point is at the lower limit	Comply with the design requirements	

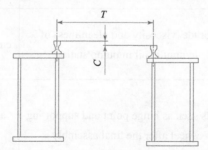

Figure 11.2.4-1　Height difference of trolley track

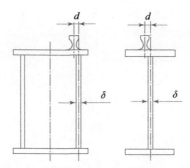

Figure 11.2.4-2   Track and center of track beam web

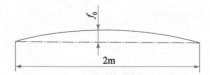

Figure 11.2.4-3   Local bending diagram

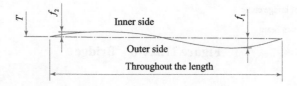

Figure 11.2.4-4   Diagram for local bending throughout the length

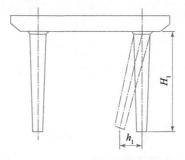

Figure 11.2.4-5   Gantry leg inclination

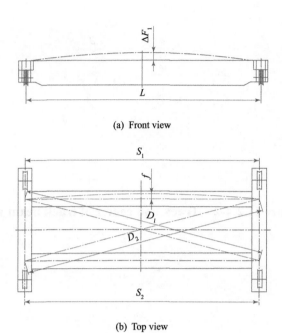

(a) Front view

(b) Top view

Key

L   bridge span

D_1, D_2 diagonal length of the bridge

S_1, S_2 crane span of bridge crane

ΔF_1 camber of main beam

Figure 11.2.4-6　Bridge

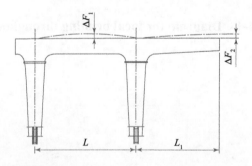

Key

L   bridge span

ΔF_1 camber of main beam

ΔF_2 effective camber of main beam on the cantilever end

(a) Front view

Figure 11.2.4-7 (I)　Gantry

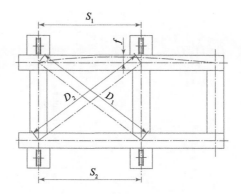

Key

D_1, D_2 diagonal length of bridge

S_1, S_2 crane span of bridge crane

f horizontal bending of main beam

(b) Top view

Figure 11.2.4-7 (II) Gantry

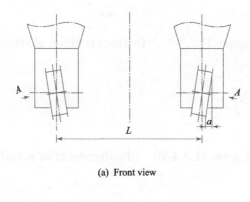

(a) Front view

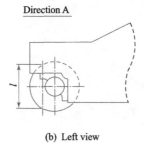

(b) Left view

Figure 11.2.4-8 Vertical inclination of wheel

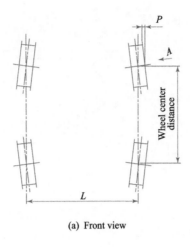

(a) Front view

(b) Left view

Figure 11.2.4-9　Horizontal offset of wheel

Figure 11.2.4-10　Misalignment of wheel

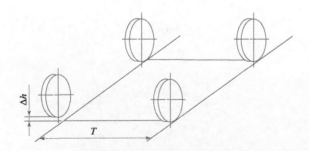

Figure 11.2.4-11　Contact between trolley driving wheel and pre-assembled track

11.2.5 The quality evaluation criteria for unit works for installation of electrical equipment shall be in accordance with Table 11.2.5.

Table 11.2.5 Quality evaluation criteria for unit works for installation of electrical equipment

No.	Testing item		Quality criteria	
			Qualified	Excellent
Δ1	Installation perpendicularity of electrical panel and cabinet (within 1 m), mm		≤ 1.5	≤ 1.0
2	Horizontal installation deviation of electrical panel and cabinet, mm	Top of two adjacent panels	≤ 2.0	≤ 1.5
		Top of panels in a row	≤ 5.0	≤ 4.0
3	Installation deviation of panel surface of electrical panel and cabinet, mm	Top edges of two adjacent panels	≤ 1.0	≤ 1.0
		Panel dial in a row	≤ 5.0	≤ 4.0
4	Seam between installation panels of electrical panel and cabinet, mm		≤ 2.0	≤ 1.5
5	Deviation of bending degree of the cable sleeve after bending, which is not greater than the outer diameter D of the sleeve		10 %D	8 %D
Δ6	Insulation resistance of resistor at normal temperature, MΩ		≥ 1	
Δ7	Insulation resistance of a complete set of electrical equipment such as electrical panel, cabinet and connecting cable, MΩ	Primary and secondary circuits in non-humid area	≥ 1	
		Primary and secondary circuits in humid area	≥ 0.5	
Δ8	Insulation resistance at normal temperature of motors with a rated voltage below 1,000 V, MΩ		≥ 0.5	
9	Installation of resistor box	Connection between resistance elements	Bare conductor shall be adopted, and the connection shall be firm and reliable	
		Lead wire splint or bolt	A mark corresponding to the wiring diagram of equipment shall be available	

Table 11.2.5 *(continued)*

No.	Testing item		Quality criteria	
			Qualified	Excellent
9	Installation of resistor box	Where the incoming or outgoing cable of the resistor box or cabinet is connected to the resistor binding post	Beeswax insulation sleeve shall be adopted	
		Stacking of resistor boxes	The number of stacks shall not exceed 4, otherwise be fixed with brackets, with heat dissipation measures available	
		Outdoor resistor box	Rainproof measures shall be available	
10	Suspending cable trolley	Installation of suspending cable trolley device	Comply with the current national standard GB 50256, *Code for Construction and Acceptance of Electric Device of Crane Electrical Equipment Installation Engineering*	
		Cable slideway	Flat, straight and smooth; the suspending cable trolley device shall be able to move agilely on the slideway, without jumping and jamming	
		Installation position of drawbar	Limit position not affecting the travel of trolley to the end when all cable trolleys return to the starting position	
		Cable clamp of suspending device	It shall be fixed with cable reliably; the distance between cable clamps should not be greater than 5 m; The free length of cable shall be 15 % to 20 % longer than the moving distance of the crane, and traction wire ropes and buffers should be installed between cable trolleys	

Table 11.2.5 *(continued)*

No.	Testing item		Quality criteria	
			Qualified	Excellent
11	Slide conductor	Installation of lifting, power supply, current collector and slide conductor	The spacing shall be uniform, about 1.5 m; The wiring of power supply shall be firm, and the current collector shall move up and down freely; Compensation measures shall be taken for slide conductor; when dispersion compensation is adopted, a clearance of 10 mm to 20 mm shall be left between conductors; when the temperature difference is large or the total length exceeds 200 m, the thermal expansion segment compensation method shall be adopted	
		Horizontal or vertical deviation between slide conductors	≤ 10 mm	
		Parallelism between the center line of slide conductor and the center line of hoist travel track	≤ 10 mm	
		Installation of slide conductor	Compensation measures shall be taken for slide conductor; when dispersion compensation is adopted, a clearance of 10 mm to 20 mm shall be left between conductors	

12 Installation of Hydraulic Hoist

12.1 Quality Evaluation Conditions

12.1.1 The quality evaluation conditions for installation of hydraulic hoist shall be in accordance with Articles 10.1.1 through 10.1.4 of this standard.

12.1.2 The hydraulic pipelines shall not be subjected to butt welding on site; the seals in the heat affected zone shall be removed when the pipelines are welded. The welding of pipelines shall comply with the current sector standard JB/T 5000.11, *Heavy Mechanical General Techniques and Standards—Part 11 Attached Piping*; the welds of stainless steel pipes shall be subjected to pickling and passivation after welding, and the surface after pickling shall be free of scars in uneven colors.

12.2 Evaluation of Unit Works for Installation of Hydraulic Hoist

12.2.1 Upon the completion of installation, the hydraulic hoist shall be subjected to on-site test and acceptance in accordance with the current sector standard NB/T 35051, *Code for Manufacture Erection and Acceptance of Gate Hoists in Hydropower Projects*.

12.2.2 The quality evaluation criteria for the installation of hydraulic hoist shall be in accordance with Table 12.2.2.

Table 12.2.2 Quality evaluation criteria for installation of hydraulic hoist

No.	Testing item	Quality criteria	
		Qualified	Excellent
Δ1	Transverse and longitudinal center lines of rack, mm	≤ 2.0	≤ 1.5
2	Rack elevation, mm	±5.0	±4.0
	Elevation and chainage of embedded hinged support of dual-lifting-point radial gate hoist, mm	±2	±1.5
Δ3	Clearance of joint surface between rack steel beam and thrust support, mm	≤ 0.05	≤ 0.05
4	Local clearance of joint surface between rack steel beam and thrust support	The clearance shall not be greater than 0.1 mm, the depth shall not be greater than 1/3 of the width of joint surface, and the cumulative length shall not be greater than 20 % of the circumference	

Table 12.2.2 *(continued)*

No.	Testing item		Quality criteria	
			Qualified	Excellent
Δ5	Levelness of thrust support top (within 1 m), mm		≤ 0.2	≤ 0.2
Δ6	Local perpendicularity of piston rod before connecting with the gate (within 1 m), mm		≤ 0.5	≤ 0.4
Δ7	Perpendicularity of full-length piston rod		Not greater than 1/4000 of the length of piston rod	
8	Pipe installation	Hydraulic pipe	The pipe bending shall comply with the current sector standard JB/T 5000.11, *Heavy Mechanical General Techniques and Standards—Part 11: Attached Piping*; After piping on the construction site, the pipeline shall be subjected to systematic flushing which shall comply with the current national standard GB/T 6996, *Common Technical Condition of Heavy Machinery Hydraulic System*; The hydraulic pipe shall be short, less curved and neat. The bending angle shall not be less than 90°. The high- and low-pressure pipes shall be distinguished in color. The pipe spacing shall meet the installation, operation and maintenance requirements of pipelines, valves, flanges, etc. The distance between the outer contours of adjacent pipelines shall not be less than 10 mm; Stainless steel gaskets or plastic and rubber gaskets without chloride ion shall be used and shall not be in direct contact with carbon steel pipe clamps. The two straight sides at the bend shall be fixed by pipe clamps; the pipe shall be firmly supported at its end and in its length direction by pipe clamps, and the spacing between pipe clamps shall comply with the current sector standard NB/T 35020, *Design Specifications for Hydraulic Hoist in Hydropower and Water Resource Projects*	

Table 12.2.2 *(continued)*

No.	Testing item		Quality criteria	
			Qualified	Excellent
9	Pipe installation	Overall circular flushing of pipes	When a special hydraulic pump station is used, the hydraulic system and hydraulic cylinder circuit of hydraulic hoist shall be cut off; when be flushed, the flow velocity in the pipe shall reach the turbulent state, the filter precision shall not be less than 10 μm, and the flushing time shall ensure that the pollution degree of flushing liquid meets the design requirements	
10	Hose installation	High-pressure hose	Comply with the current national standard GB/T 3683.1, *Rubber Hoses and Hose Assemblies—Wire-braid-reinforced Hydraulic Types—Specification—Part 1: Oil-based Fluid Applications*. The hose, when used, shall not be tightened or twisted, and shall not be rubbed against other objects when moving. The length of hose from the joint to the bend shall not be less than 6 times the outer diameter of the hose, and the bending radius shall not be less than 10 times the outer diameter of the hose	
11	Hydraulic oil		Comply with the current sector standard NB/T 35051, *Code for Manufacture Erection and Acceptance of Gate Hoists in Hydropower Projects*	
12	Field test		Comply with Article 12.2.1 of this standard	
13	Joint commissioning without water		Comply with Article 12.2.1 of this standard	
14	Joint commissioning with water		Parameters of hydraulic hoist such as pressure, opening/closing speed and stroke shall meet the design requirements, all signals and displays shall be accurate, and the protection function shall be safe and reliable	

12.2.3 The quality evaluation criteria for unit works for installation of electrical equipment shall be in accordance with Article 10.2.5 of this standard.

13 Installation of Screw Rod Hoist

13.1 Quality Evaluation Conditions

13.1.1 The work breakdown for installation of screw rod hoist shall be in accordance with Article 10.1.1 of this standard.

13.1.2 The work on screw rod hoist before delivery shall be in accordance with Article 10.1.2 of this standard.

13.1.3 The technical requirements for installation, welding, surface corrosion protection and inspection of screw rod hoist shall be in accordance with Article 10.1.3 of this standard.

13.1.4 The assembly and testing in factory and the installation and trial run on site of screw rod hoist equipment shall be in accordance with Article 10.1.4 of this standard.

13.2 Evaluation of Unit Works for Installation of Screw Rod Hoist

13.2.1 The screw rod hoist should be assembled in factory, and may be assembled on site if agreed upon when the screw is too long. However, before delivery, the nut shall be screwed in the full stroke of screw rod without jamming.

13.2.2 The no-load test and load test for screw rod hoist shall comply with the current sector standard NB/T 35051, *Code for Manufacture Erection and Acceptance of Gate Hoists in Hydropower Projects*.

13.2.3 The quality evaluation criteria for installation of screw rod hoist shall be in accordance with Table 13.2.3.

Table 13.2.3 Quality evaluation criteria for installation of screw rod hoist (mm)

No.	Testing item	Quality criteria	
		Qualified	Excellent
Δ1	Vertical and horizontal center lines of hoist body	±2.0	±1.5
2	Base seat elevation	±5.0	±4.0
Δ3	Levelness of upper plane of base plate	≤ 0.5/1000	≤ 0.4/1000
4	Perpendicularity of screw before being connected to gate	≤ 0.2/1000	

Table 13.2.3 *(continued)*

No.	Testing item	Quality criteria	
		Qualified	Excellent
5	Local clearance of contact surface between hoist body and base plate	Contact tight, the local clearance is not greater than 0.5 and the contact area is not less than 70 % of the total area	
Δ6	No-load test	Comply with the current sector standard NB/T 35051, *Code for Manufacture Erection and Acceptance of Gate Hoists in Hydropower Projects*	
Δ7	Load test	Comply with the current sector standard NB/T 35051, *Code for Manufacture Erection and Acceptance of Gate Hoists in Hydropower Projects*	

13.2.4 The quality evaluation criteria for unit works for installation of electrical equipment shall be in accordance with Article 10.2.5 of this standard.

Appendix A Quality Evaluation Forms for Unit Works

A.0.1 The format of quality inspection record for unit works should comply with Table A.0.1.

Table A.0.1 Quality inspection record for unit works

Partioned project		Unit works				Position		
Contractor		Work quantities (t)				Commencement date		

No.	Testing item	Permissible deviation (mm)		Measured value (mm)							
		Excellent	Qualified	1	2	3	4	5	6	7	8

Total	Total number of inspection items:____, wherein, number of qualified works:____ and number of excellent works:____
Conclusion	

Representative of supervisor (owner)		Technical director of contractor				Tested by		

A.0.2 The format of quality evaluation form for unit works should comply with Table A.0.2.

Table A.0.2 Quality evaluation form for unit works

Partioned project		Unit works		Position		
Contractor		Work quantities (t)		Commencement and completion dates		
No.	Item	Main items		General items		
		Excellent (pcs)	Qualified (pcs)	Excellent (pcs)	Qualified (pcs)	
	Total					
Total number of inspection items:____, wherein, number of excellent works:____ and excellent rate:____%						
Conclusion			Evaluation grade			
Representative of supervisor (owner)			Technical director of contractor			

Explanation of Wording in This Standard

1   Words used for different degrees of strictness are explained as follows in order to mark the differences in implementing the requirements of this standard.

   1) Words denoting a very strict or mandatory requirement:

      "Must" is used for affirmation; "must not" for negation;

   2) Words denoting a strict requirement under normal conditions:

      "Shall" is used for affirmation; "shall not" for negation;

   3) Words denoting a permission of a slight choice or an indication of the most suitable choice when conditions permit:

      "Should" is used for affirmation; "should not" for negation;

   4) "May" is used to express the option available, sometimes with the conditional permit.

2   "Shall comply with…" or "shall meet the requirements of…" is used in this standard to indicate that it is necessary to comply with the requirements stipulated in other relative standards and codes.

List of Quoted Standards

GB/T 1182,   *Geometrical Product Specifications (GPS)—Geometrical Tolerancing—Tolerances of Form, Orientation, Location and Run-out*

GB/T 3323,   *Radiographic Examination of Fusion Welded Joints in Metallic Materials*

GB/T 3683.1, *Rubber Hoses and Hose Assemblies—Wire-braid-reinforced Hydraulic Types—Specification—Part 1: Oil-based Fluid Applications*

JB/T 6996,   *Common Technical Condition of Heavy Machinery Hydraulic System*

GB/T 8923.1, *Preparation of Steel Substrates Before Application of Paints and Related Products—Visual Assessment of Surface Cleanliness—Part 1: Rust Grades and Preparation Grades of Uncoated Steel Substrates and of Steel Substrates after Overall Removal of Previous Coatings*

GB/T 8923.2, *Preparation of Steel Substrates Before Application of Paints and Related Products—Visual Assessment of Surface Cleanliness—Part 2: Preparation Grades of Previously Coated Steel Substrates after Localized Removal of Previous Coatings*

GB/T 11345, *Non-destructive Testing of Welds—Ultrasonic Testing—Techniques, Testing Levels, and Assessment*

GB 50256,   *Code for Construction and Acceptance of Electric Device of Crane Electrical Equipment Installation Engineering*

GB 50766,   *Code for Manufacture Installation and Acceptance of Steel Penstocks in Hydroelectric and Hydraulic Engineering*

NB/T 35020, *Design Specifications for Hydraulic Hoist in Hydropower and Water Resources Projects*

NB/T 35045, *Code for Manufacture Installation and Acceptance of Steel Gates in Hydropower Engineering*

NB/T 35051, *Code for Manufacture Erection and Acceptance of Gate Hoists in Hydropower Projects*

NB/T 47013.3, *Nondestructive Testing of Pressure Equipments—Part 3:*

Ultrasonic Testing

NB/T 47013.4, *Nondestructive Testing of Pressure Equipments—Part 4: Magnetic Particle Testing*

NB/T 47013.5, *Nondestructive Testing of Pressure Equipments—Part 5: Penetrant Testing*

NB/T 47013.10, *Nondestructive Testing of Pressure Equipments—Part 10: Ultrasonic Time of Flight Diffraction Technique*

DL/T 330, *Ultrasonic Time of Flight Diffraction Technique in Welded Joints of Hydroelectric or Hydraulic Engineering Steed Structure and Equipment*

SL 105, *Specifications for Anticorrosion of Hydraulic Steel Structure*

JB/T 5000.11, *Heavy Mechanical General Techniques and Standards—Part 11: Attached Piping*

GB/T 29712, *Non-destructive Testing of Welds—Ultrasonic Testing—Acceptance Levels*